Designed and produced by
Aladdin Books Ltd
70 Old Compton Street
London W1

Design David West
Children's Book Design
Editorial Planning Clark Robinson Limited
Editor Bibby Whittaker
Researcher Cecilia Weston-Baker
Illustrated by Ron Hayward Associates
and Aziz A. Khan

EDITORIAL PANEL

The author, Linda Gamlin, has degrees in Biochemistry and Applied Biology, and has contributed to several encyclopedias.

The educational consultant, Peter Thwaites, is Head of Geography at Windlesham House School in Sussex.

The editorial consultant, John Clark, has contributed to many information and reference books.

First published in the
United States in 1988 by
Gloucester Press
387 Park Avenue South
New York, NY 10016

ISBN 0-531-17119-1

Library of Congress Catalog
Card Number: 88-50509

Printed in Belgium

TODAY'S WORLD

ORIGINS OF LIFE

LINDA GAMLIN

GLOUCESTER PRESS
New York · London · Toronto · Sydney

CONTENTS

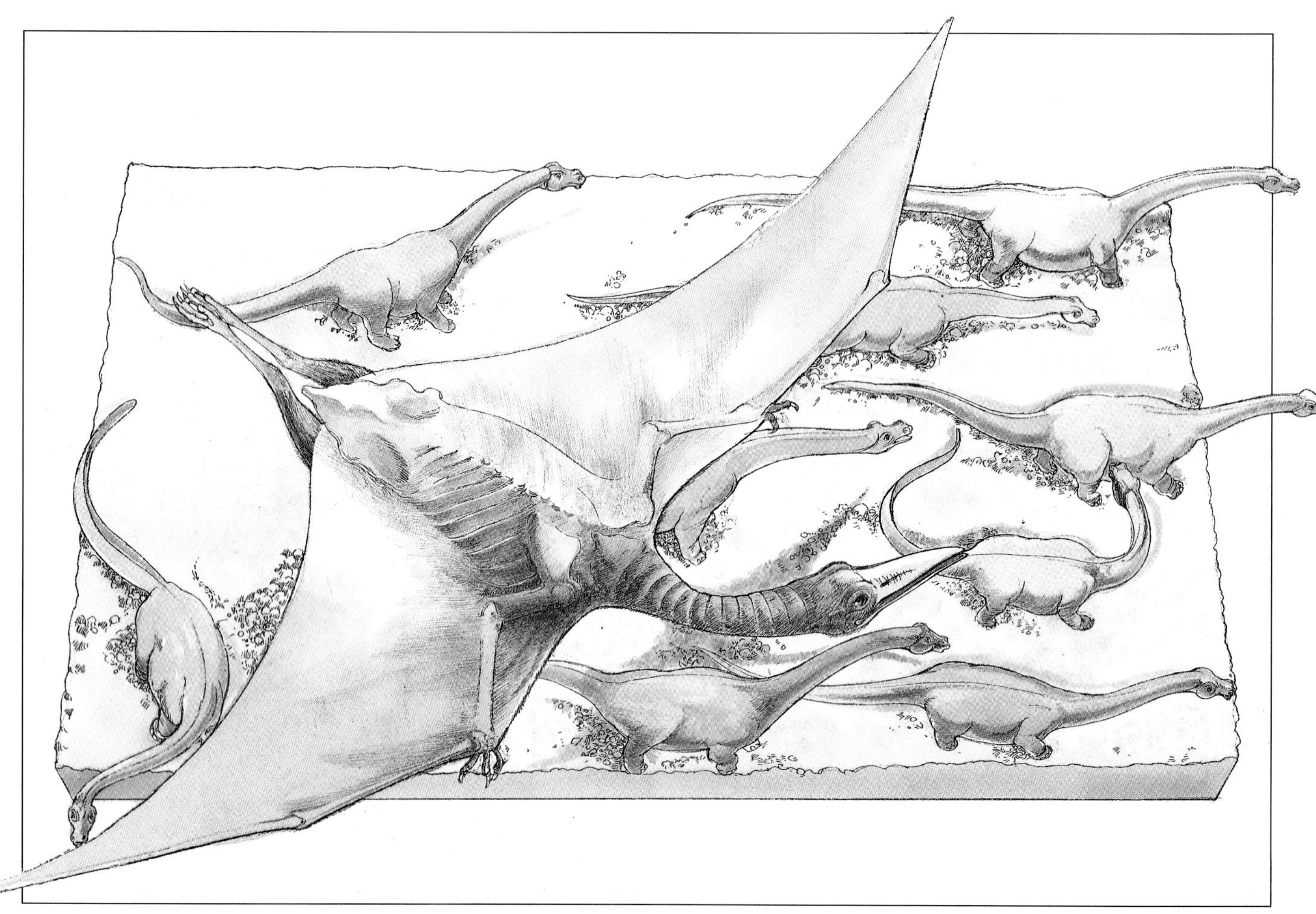

The front cover photograph shows a computerized image of the structure of DNA.

INTRODUCTION

Life began on Earth with very simple organisms, which probably lived in the sea. They were single-celled (unicellular) microscopic creatures resembling bacteria. This book tells the story of how they gradually evolved into more complicated organisms. The first many-celled forms were seaweeds and sponges, followed by jellyfish, worms and starfish. There were also jointed animals – the arthropods, such as lobsters, spiders and insects – as well as shellfish and other molluscs.

The first animals with backbones were fish. Some of these evolved into amphibians, newt-like animals that could crawl on land. These gave rise in turn to reptiles, from fierce crocodiles which remained in the water to snakes and lizards that lived on the land. The largest of all were the dinosaurs. One branch of small dinosaurs developed feathers and evolved into birds. Another group gave rise to mammals, which remain the most highly evolved animals living on Earth today.

A chimpanzee, one of the most advanced mammals, shaping a twig to use as a "tool" to probe for termites

HOW LIFE BEGAN

The Earth is about 5 billion years old. Life probably began 3.5 to 4 billion years ago. The oldest known fossils are cyanobacteria – relatives of bacteria that can use sunlight for making food, as green plants do. They date from about 3.2 billion years ago. But they were probably not the first form of life. Other types of bacteria-like cells came before them.

Life began on Earth in conditions that were very different from those we have today. The air contained little oxygen, and there were many erupting volcanoes and violent electrical storms. The lightning caused simple chemicals in the atmosphere to combine and form more complex chemical compounds – the basic ingredients of life. These included molecules such as nucleic acids like DNA, and amino acids – the building blocks of proteins, which in turn are the building blocks of living matter. Once such substances were available, the first cells began to form.

All living things are made of cells, from single-celled bacteria to the thousands of species of multicellular plants and animals that inhabit the Earth today. The original form of life was probably a single-celled organism resembling a bacterium.

From molecules to cells

There are 92 chemical elements in nature, from the lightest hydrogen to the heaviest uranium. Elements can combine to form simple molecules. For example, hydrogen combines with two atoms of oxygen to form a molecule of water. Complex molecules, such as amino acids, are formed by several elements combining. Two types of complex molecules form living things: proteins (made up of different combinations of amino acids) and nucleic acids (consisting of strings of units called nucleotides). All life, from bacteria to oak trees to humans, is made up of these same chemical building blocks. Over a long period of time, the "primeval soup" of the Earth's seas became rich in complex molecules such as amino acids. These began to link together to form long-chain molecules – the first proteins. In time, these molecules grouped together with other long-chain molecules to form the first primitive bacteria-like cells.

Oxygen and life

The early living cells probably survived by eating the complex molecules in the primeval soup around them. As time went by, this supply of food began to run out, and some cells adapted by creating their own food using light energy from the Sun. This process is photosynthesis, and the cells that "created" it were types of bacteria. Among the most advanced of them were the cyanobacteria, and they developed a special version of photosynthesis that was more efficient. The new process had an important by-product – oxygen gas. As a result, over millions of years, oxygen built up in the atmosphere. One sign of this is bands of rust-colored iron oxide in old rocks, where the oxygen first reacted with iron.

Bands of iron oxide in rock

The first cells

All cells are surrounded by membranes, made of fat-containing molecules. In water, such molecules may form double layers, and this is how the first membranes probably came about. If they developed around groups of other complex molecules in the primeval soup, they could have produced cell-like globules. In time these could have become simple cells, like those of present-day bacteria.

The importance of DNA

One key step in the evolution of life was the development of DNA (deoxyribonucleic acid). In modern living things, DNA forms the genes which carry all the information needed to make proteins. The proteins make everything else in the cell – including more DNA. The diagram on pages 28-29 illustrates this close relationship between proteins and DNA.

A bacterial cell

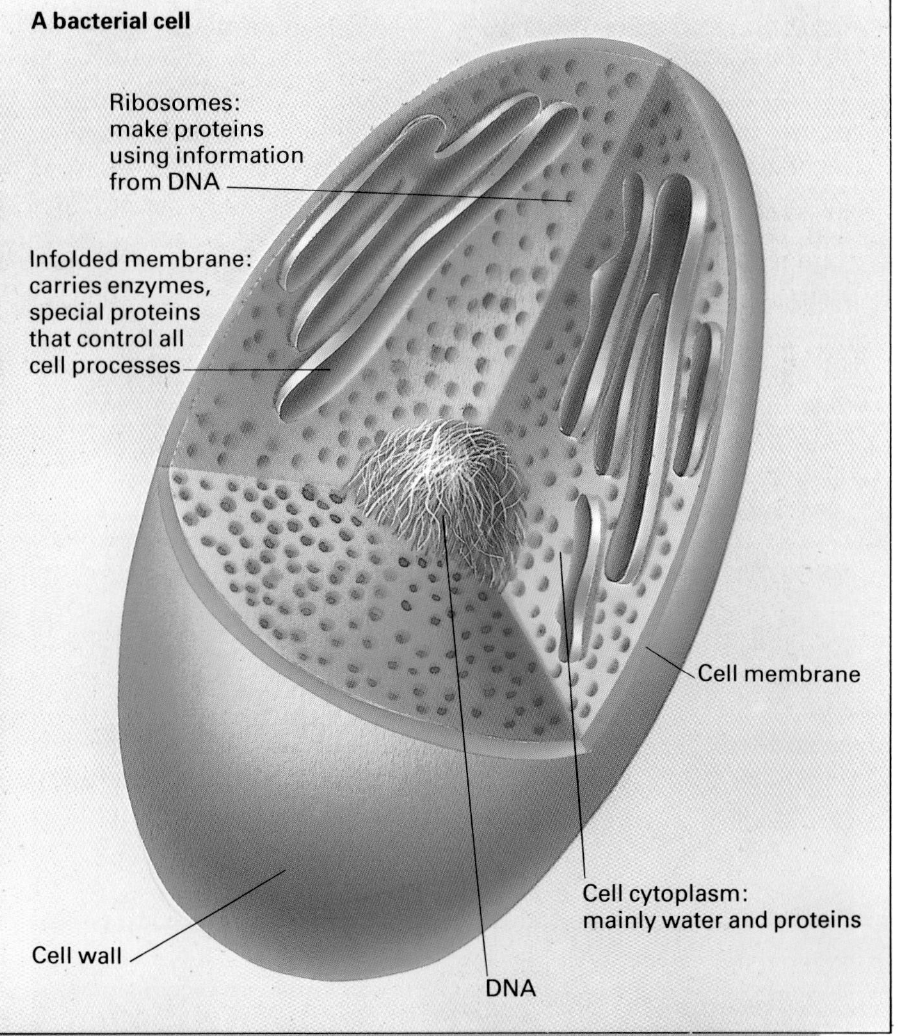

THE FIRST PLANTS

Bacteria are so small that they can be seen only with the aid of a microscope. Most are 0.001 to 0.005 millimeters across. They have cells with a very simple structure. The cells of plants and animals are more complex and measure between 0.01 and 0.05 millimeters across.

The amount of oxygen in the atmosphere gradually increased with the greater number of cyanobacteria. At first it probably killed off many of the bacteria, but gradually they adapted to use the oxygen for their own purposes. The development of different types of bacteria led to the evolution of other organisms that obtained their energy simply by feeding off more primitive ones. The next important stage occurred when more complex cells acquired the ability to photosynthesize through the development of chloroplasts (see below). These were the first true plant cells.

Simple to complex cells

The cells of plants and animals are much more complicated than those of bacteria. Plants, for example, have tiny egg-shaped structures called chloroplasts inside their cells. By separating them from the rest of the cell, scientists have shown that they do the work of photosynthesis – using sunlight to make food. What is most striking about chloroplasts is how much they resemble cyanobacteria – the single-celled creatures that originally pumped oxygen into the air. It is now believed that chloroplasts are the descendants of cyanobacteria that made a home inside larger cells many millions of years ago.

Plant and animal cells include other egg-shaped structures known as mitochondria. These are responsible for the reactions in which food is broken down and oxygen is used to produce energy. Like chloroplasts, mitochondria are probably descended from bacteria that had learned to cope with oxygen, and which then began living inside larger cells that had "swallowed" them.

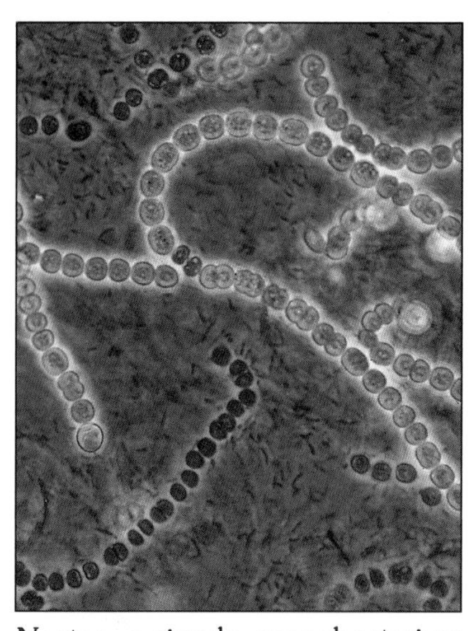

Nostoc, a simple cyanobacterium

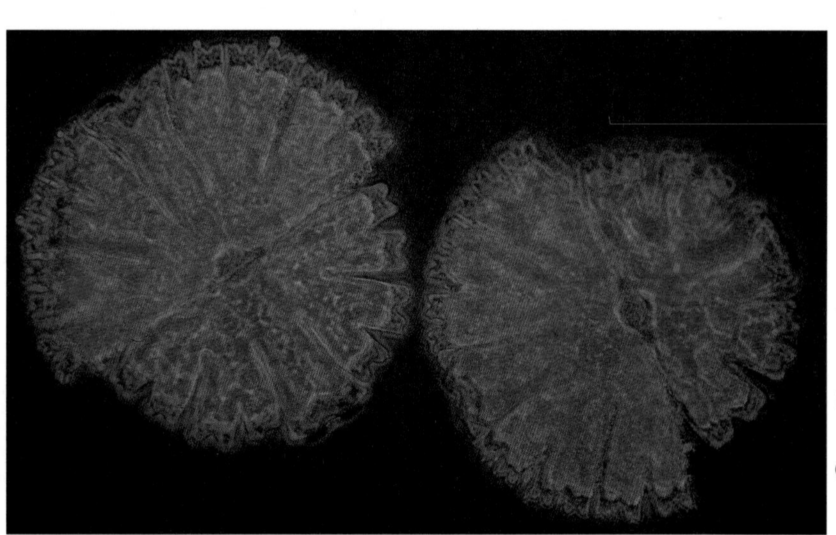

Single-celled desmids have a cellulose cell wall.

PLANT CELL

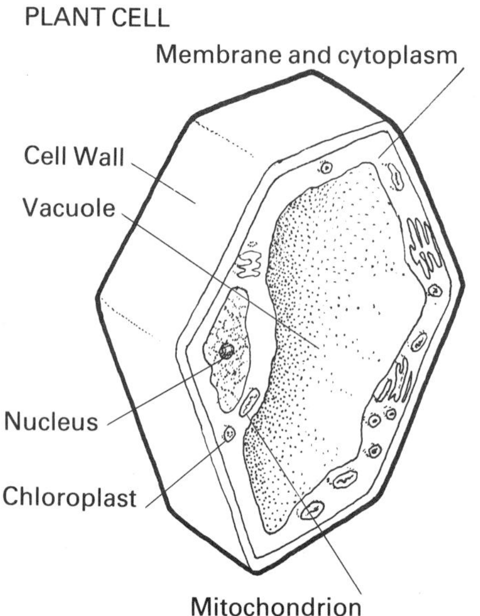

Plant evolution

The first plants were single-celled organisms formed when cyanobacteria took refuge inside larger cells. Those larger cells already contained mitochondria, descended from other bacteria. From their point of view, there was a lot to gain by giving the bacteria a home. The mitochondria helped them to cope with oxygen, and the chloroplasts made food for them.

These ancestral plants probably lived at the surfaces of seas and lakes, as many of their descendants – the unicellular algae – do today. In time, the single-celled plants began to evolve into many-celled forms, by dividing into two without separating. Some formed balls of cells, others hollow cylinders, and some formed strings of cells. Many organisms with these arrangements (shown below) live today as pondweeds and seaweeds.

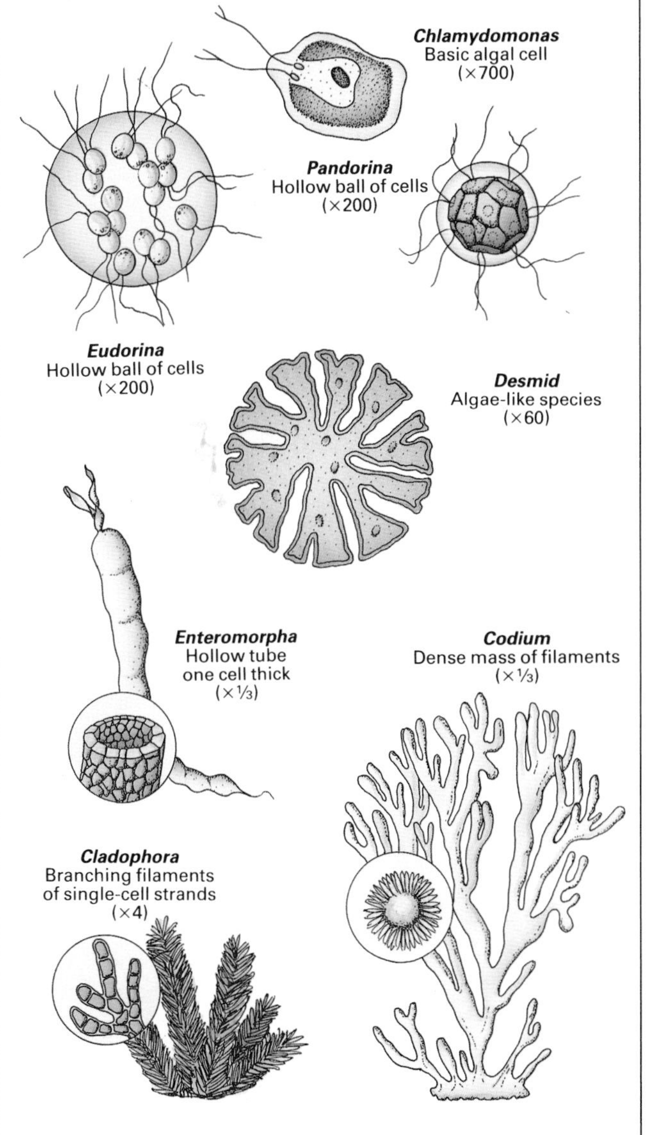

On to land

For millions of years, algae were the only forms of plant life. Like today's algae, they were largely confined to water because they would dry out on land and because they had no supporting structure to hold up their leaf-like fronds. To begin with, therefore, there were no plants on land. Then a small alga began to grow around the edges of ponds, just out of the water. In time, it evolved a semi-waterproof covering, and developed root-like structures to draw up water from the soil. These pioneering algae gave rise to the mosses, simple plants which still need damp places in which to grow. Later on, more advanced groups of plants such as ferns appeared. These had a waterproof covering on their leaves and were the first plants to develop roots and woody stems, allowing them to grow taller.

Mosses need a layer of moisture to reproduce.

The cycad is an ancient primitive plant.

THE FIRST ANIMALS

Precambrian Era: 4 billion-590 million years ago. **1,000 mya:** complex single cells appear. **700 mya:** multicellular animals appear. There are fossils of complete animals preserved in sediments and rocks from this period, including some strange looking jellyfish floating in the seas and feeding on small floating animals. **Cambrian Era:** 590-500 mya. Shelled animals.

Animals, like plants, evolved from complex single cells that built themselves into multicellular organisms by dividing without separating. But the single cells that gave rise to animals were different from those that became plants – they contained mitochondria but no chloroplasts, so they could not photosynthesize. Instead, they fed on other primitive plants and bacteria. Distant relatives of those single-celled animal ancestors are still alive today – the protozoa.

A later stage in the development of multicellular organisms was one in which individual cells began to perform separate tasks like digestion or movement. This division of jobs among cells led to the evolution of many more complex organisms.

Single to multicellular

Protozoa

In the early history of life in the sea, numerous single-celled plants and animals would have come into existence. Today's protozoa descended from these early life forms and can be regarded as the simplest animal-like organisms. A typical example is the amoeba, which consists of a single cell containing jelly-like protoplasm. It moves by extending "false feet" (pseudopodia) and dragging itself along. It eats by forming a pouch in its outer membrane and encircling a particle of food to engulf it. Some amoebae can be parasites, living in humans and animals, where they cause diseases such as dysentery.

Multicellular creatures

As the numbers of single-celled organisms increased, so too would the variety of creatures that fed on them. The earliest many-celled animals were creatures such as jellyfish and worms. Animals like these do not easily form fossils because there are no hard parts to their bodies. But in very fine sediments, they did leave impressions of their burrows and tracks that have been preserved. Such sediments, dating from 700 million years ago, have survived in southern Australia and a few other places. But these are the exceptions – in general there are very few fossils from this Precambrian era.

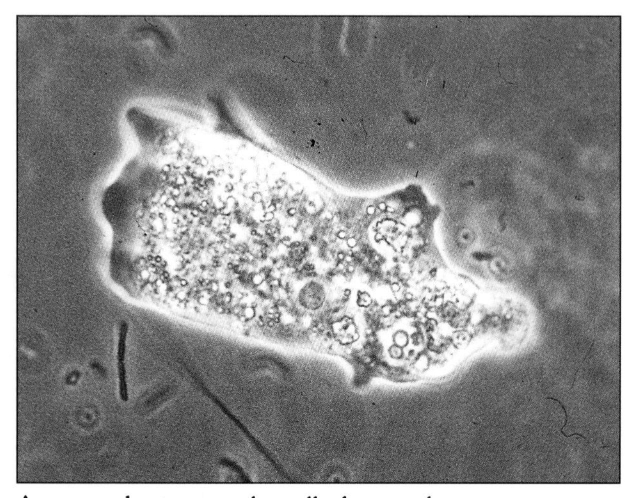

An amoeba is a single-celled animal.

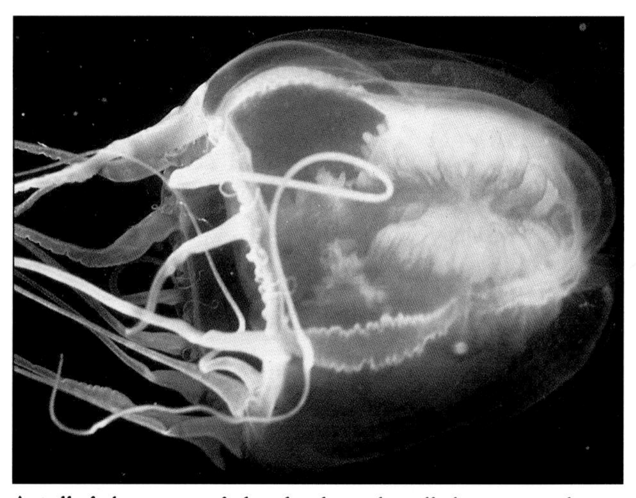

A jellyfish is a soft-bodied multicellular animal.

Shelled animals

Rocks from the beginning of the Cambrian era show great changes had taken place. Here the rocks are teeming with fossils, particularly those of trilobites (which looked a bit like modern woodlice). Also common are graptolites, which were worm-like animals that lived in chalky tubes. All these animals are now extinct, but their hard shells have remained as fossils.

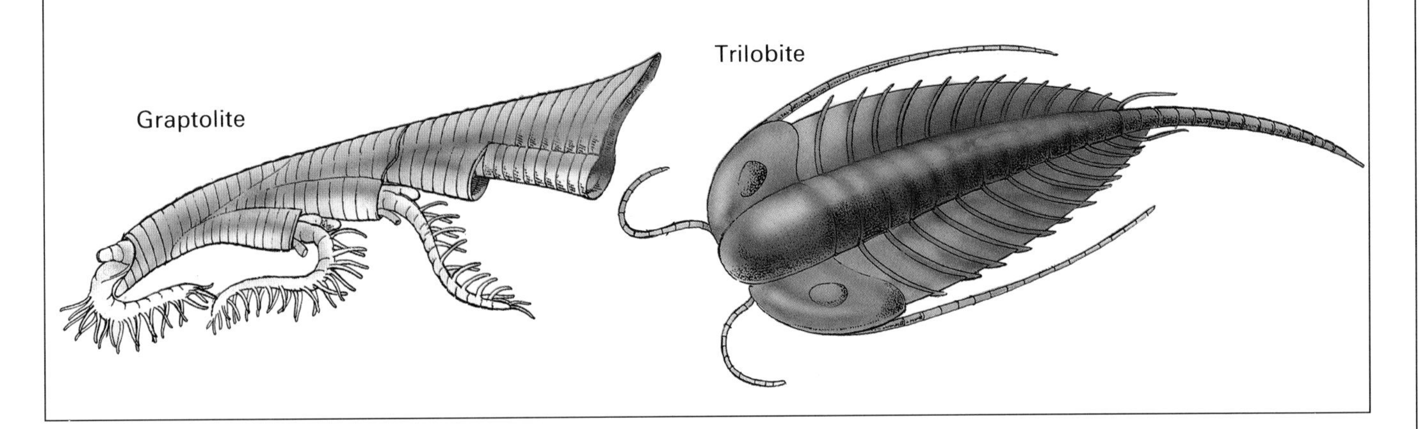

Invertebrate evolution

Many animals that lived in the Cambrian era are now extinct, but others have left living descendants. The family tree shows how today's invertebrates (animals without backbones) might have evolved from the ancestral animal cell. Most of this evolution took place during the Cambrian period, when almost all the major groups of invertebrates appeared on Earth. But evolution has not stood still for invertebrates in the 500 million years since the Cambrian ended. The invertebrate groups have produced many new sub-groups, but more important for evolution is the fact that they eventually gave rise to animals with a notochord, like the lancelet (see pg. 10), and to backboned animals (vertebrates).

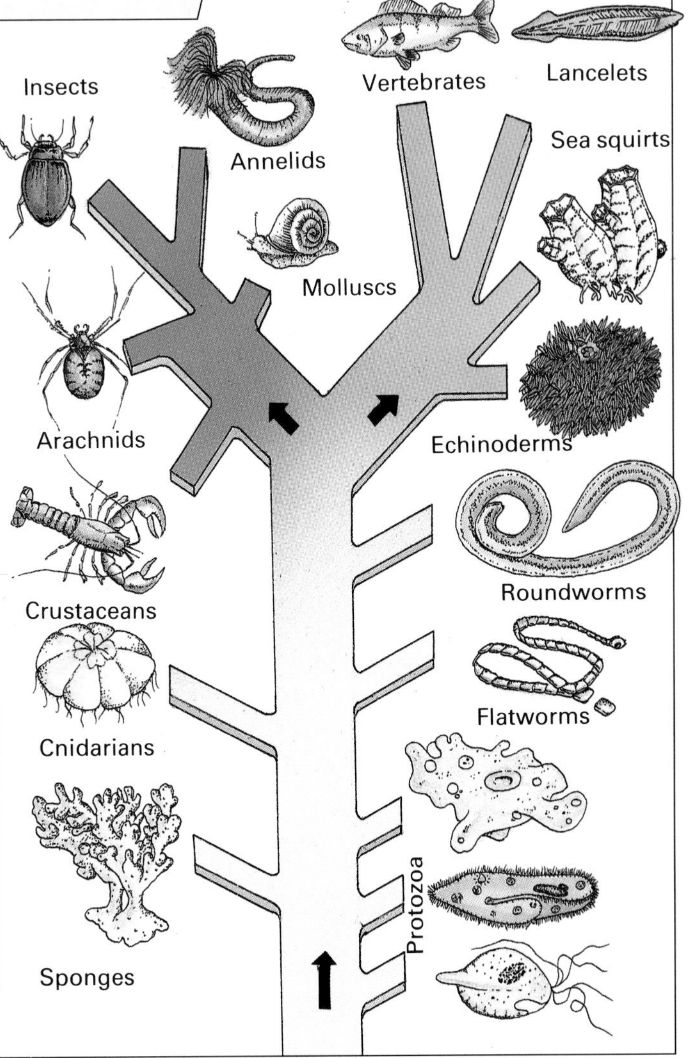

Fossil of a fly trapped in amber

THE ORIGIN OF FISH

The first fish probably evolved towards the end of the **Cambrian** period, over 500 million years ago. They flourished and diversified during the **Devonian**: 395 million to 345 million years ago. By the end of that period, all the major groups of fish had evolved.

The first vertebrates (backboned animals) were fish which evolved from invertebrates that probably resembled today's sea squirts. These bag-shaped animals live anchored to the bottom of the sea, and do not look at all like fish. But their young (larvae) are small, streamlined animals that swim about looking for a place to spend their adult lives. They have a rod of stiff, rubbery tissue – the notochord – running from head to tail. This notochord was the forerunner of the backbone, and is still found in all vertebrates – including humans – during the embryo stage.

Early fish

The earliest fish developed a bony armor by excreting calcium salts (in the same way that hard water – that which has a high content of calcium – builds up crusty deposits). The weight of this armor caused them to sink to the sea bed. The mouth was a jawless, circular opening. This type of mouth is seen in present-day jawless fish – the hagfish and lampreys. These are descended from early fish, but they have no armor. The extinct fish shown below are typical of early forms that evolved. *Coccosteus* had its head and part of its body covered with armor. *Cephalaspis* was a jawless fish with a triangular head shield and the mouth underneath. *Jamoytius* was an eel-like fish with fins along the whole length of its body. All three lived about 400 million years ago.

The lancelet

The invertebrate ancestor of fish has not survived. But another animal, the lancelet, gives us a good idea of what it probably looked like. The lancelet has muscles arranged in blocks, as in a fish. It also has a notochord. All vertebrates go through a stage in their development as embryos when they have a notochord. Later this is replaced by the spine.

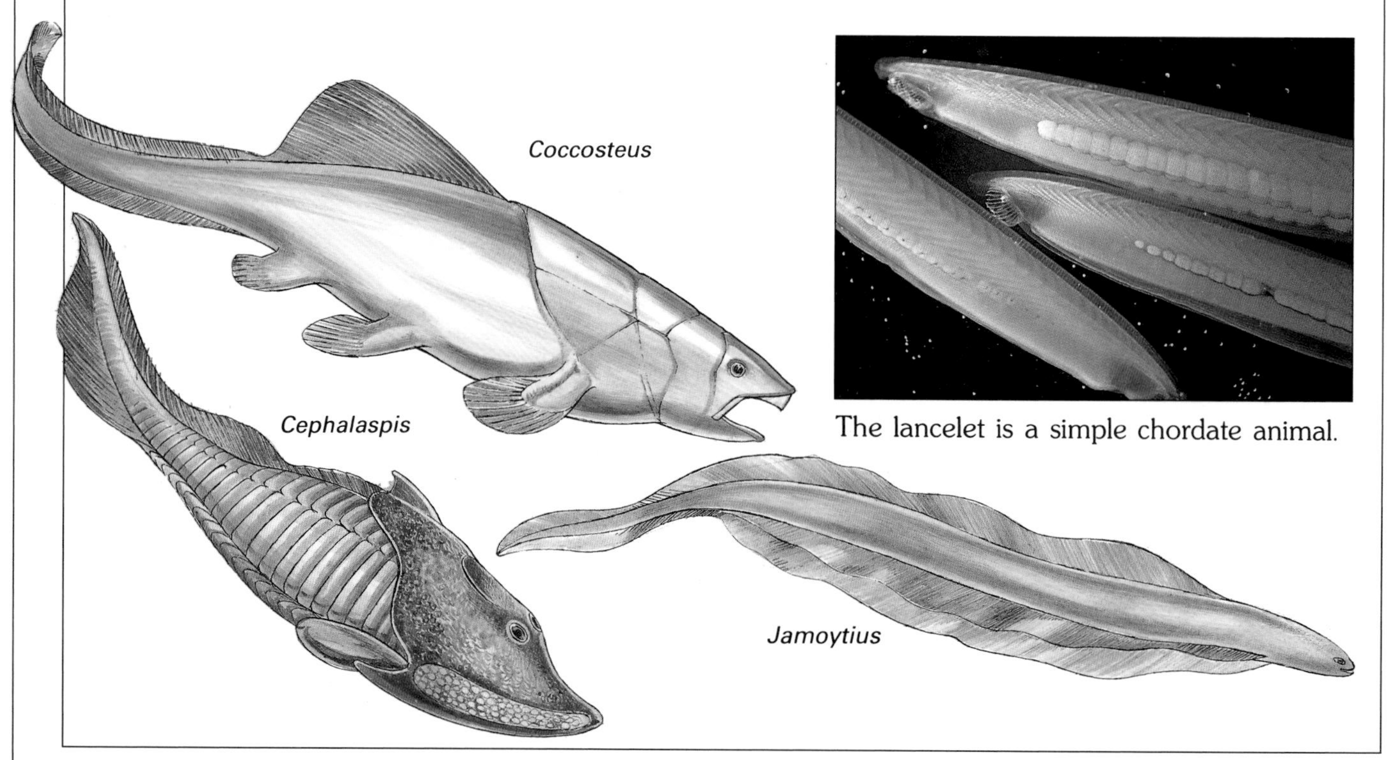

The lancelet is a simple chordate animal.

Living fossils

In almost all groups of animals there are a few species that survive almost unchanged from a much earlier age. Such species are often referred to as living fossils, although it is unlikely that they are exactly the same as their relatives that have become extinct.

The hagfish and lampreys are one such group, and the sturgeons are another. Sturgeons belong to the ray-finned fish group, but they represent a primitive group which were once very numerous in the oceans of the Earth. Now only a few species survive. Sturgeons have a bony skeleton, a shark-like tail and along their sides a row of bony, diamond-shaped plates – a reminder of the armor plating that covered most fish some 400 million years ago.

The lamprey is a primitive parasitic jawless fish.

The sturgeon is a primitive ray-finned fish.

Jaws and bones

The next important step in fish evolution was the development of jaws, which allowed fish to eat other fish. Two major groups evolved: cartilaginous fish (sharks and rays) and ray-finned fish (most of the others). Sharks and rays have a skeleton made of cartilage, a strong, rubbery material that consists of protein. In all other vertebrates, the basic skeleton consists of cartilage, but hard minerals are added to it to make bone. The ancestors of sharks and rays probably had bones, but the hardening process was lost through time. Ray-finned fish have bony skeletons, more mobile jaws than the sharks, and efficient fins for swimming.

A shark has powerful jaws and sharp teeth.

Goldfish are typical bony fish of the carp family.

FROM WATER TO LAND

The ancestors of amphibians – animals that live on land but breed in water – emerged from the water about 375 million years ago. They were the dominant land animals for 80 million years – throughout the **Devonian** and the early **Carboniferous** period that followed.

The next significant stage in evolution was the move from water on to land. The first animals to make this giant step were invertebrate arthropods – jointed animals that lived in the sea and emerged on to land to evolve into spiders, insects and millipedes. They lived among the simple plants that had already begun to grow on land, and probably fed on them as well. During this time there was an annual drought which caused many rivers to dry up. Some fish responded by coming on to land in search of food. In time, these fish evolved into amphibians.

Emergence on to land

The abundant food supply on land probably tempted fish out of the water to take advantage of it. But a successful backboned land animal needs two features that fish lack – lungs and limbs – and these are still the main differences between terrestrial (land-dwelling) amphibians and aquatic fish. Some primitive fish, such as the coelacanth, do have bony fins and may resemble the type that evolved into amphibians.

All amphibians have to return to water to lay their eggs. The young that hatch from the eggs (larvae or tadpoles) have no legs at first and they have to breathe oxygen dissolved in water using gills like a fish. As they develop into adults they change radically. This change from aquatic larvae to land-living adults is a process called metamorphosis. They grow four legs, lose their gills and develop lungs for breathing oxygen from the air. Frogs and toads also lose their tails.

A coelacanth has leg-like fins.

Emergence on to land

The ancestor of the amphibians was a fast-moving predator

It gulped air at the surface when water oxygen level was low

Teeming invertebrates on land tempted it out of the water to feed

Getting away from water

An animal that lives on land must have legs to move about and a waterproof skin to keep itself from drying out. The legs have to be stronger than a fish's fins, because air does not support an animal's weight as well as water does. Some fish, such as catfish, can emerge from water and pull themselves along on their fins. Since they cannot lift their bodies off the ground, they are unable to move very fast. Land animals must also be able to breathe air, either with lungs or other structures. Insects rely on a system of air-filled tubes that extend throughout their bodies.

A catfish can "walk" on its front fins.

Dragonflies were among the first flying insects.

Early amphibians

The earliest amphibians were fairly small, but soon evolved into larger creatures – some as much as 4m (13 ft) long. All had thick waterproof skins, much thicker than those of many modern amphibians such as frogs. The fins of their fishy ancestors had evolved into legs, but these were set at the sides of the body, rather than underneath as they are in a dog. So the early amphibians lumbered along clumsily, with their stomachs inches from the ground. They did not need to run fast because food was plentiful, and there were no land animals to prey on them.

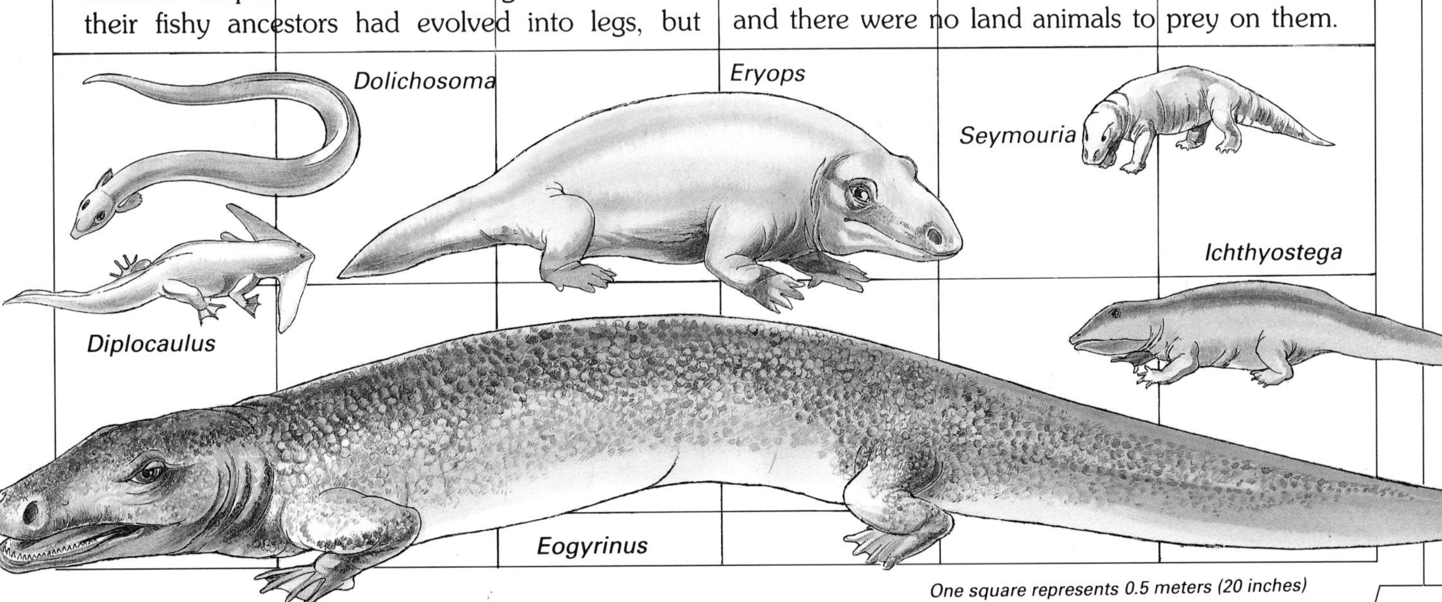

THE AGE OF REPTILES

Reptiles evolved from amphibians during the **Carboniferous** period, over 300 million years ago, and they diversified rapidly. They were the dominant land animals for over 225 million years. Some, like the dinosaurs, grew to enormous sizes. Others were ancestors of birds and mammals. The Age of Reptiles ended 65 million years ago, at the end of the **Cretaceous**, when a large number of species suddenly became extinct.

Although the amphibians had moved out on to land, they still had to return to water to lay their eggs. The eggs developed into tadpoles, like those of today's frogs and newts. To become truly independent of water, some animals produced eggs with tough, waterproof shells that could be laid on land. This enabled them to live in drier places. Their young no longer had an aquatic stage, but instead were land-dwelling, miniature versions of the adults. The amphibians that took this step evolved into reptiles – ancestors of modern lizards, snakes, crocodiles and turtles.

Like the amphibians before them, reptiles went through a very successful period when they were the dominant animals on Earth. The Age of Reptiles lasted for over 225 million years, and produced some of the most extraordinary animals the world has ever seen.

Air

One group of reptiles, the pterosaurs, took to flying through the air. They had large membranous wings and furry bodies. The presence of fur shows that they were warm-blooded like today's mammals and birds.

Land

Dinosaurs dominated the land during the latter half of the Age of Reptiles. Some, such as the sauropods, ankylosaurs and stegosaurs, were heavy, slow-moving plant-eaters. Others, such as *Tyrannosaurus*, were carnivores that ran on their hind legs and seized prey with their front legs.

Diversity

The reptiles diversified into at least eight major lines at an early stage in their evolution. Three of these have survived: turtles and tortoises, snakes and lizards, and crocodiles. A fourth group, the mammal-like reptiles, evolved into mammals, while another group evolved into birds. Three other groups – ichthyosaurs, plesiosaurs and pterosaurs – have all died out.

Of all the reptiles, dinosaurs flourished the longest and became the most diverse. Part of their success was due to their legs being positioned underneath the body rather than out at the sides. In this position they could take more weight, so dinosaurs could become larger. Later dinosaurs ran on their long hind legs.

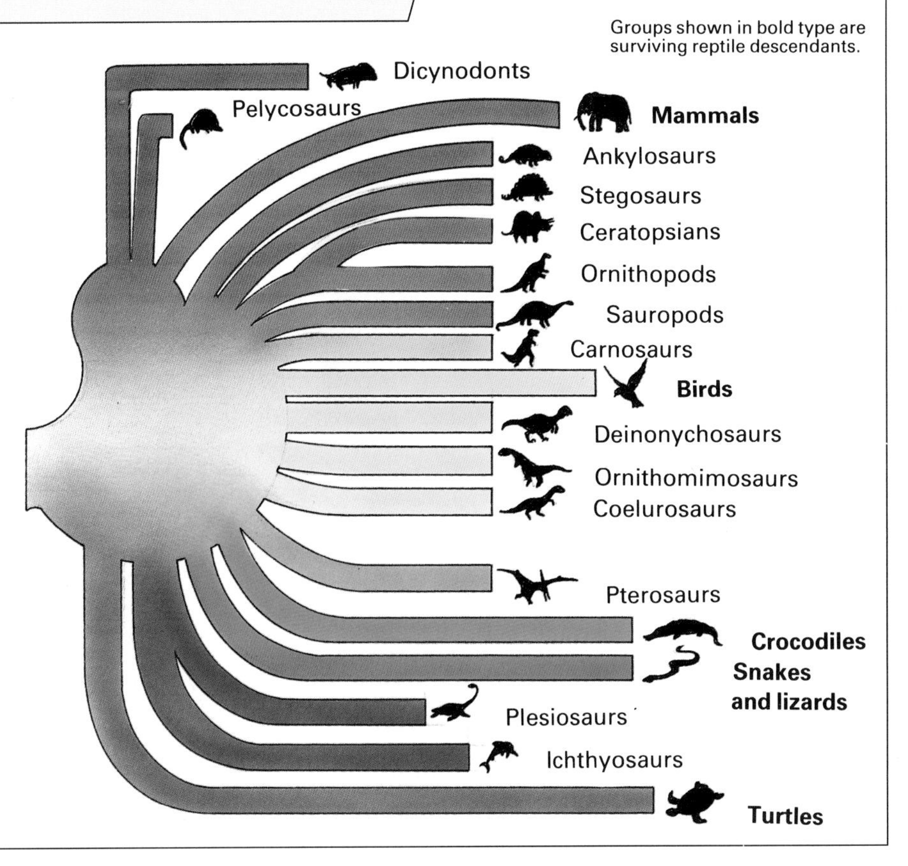

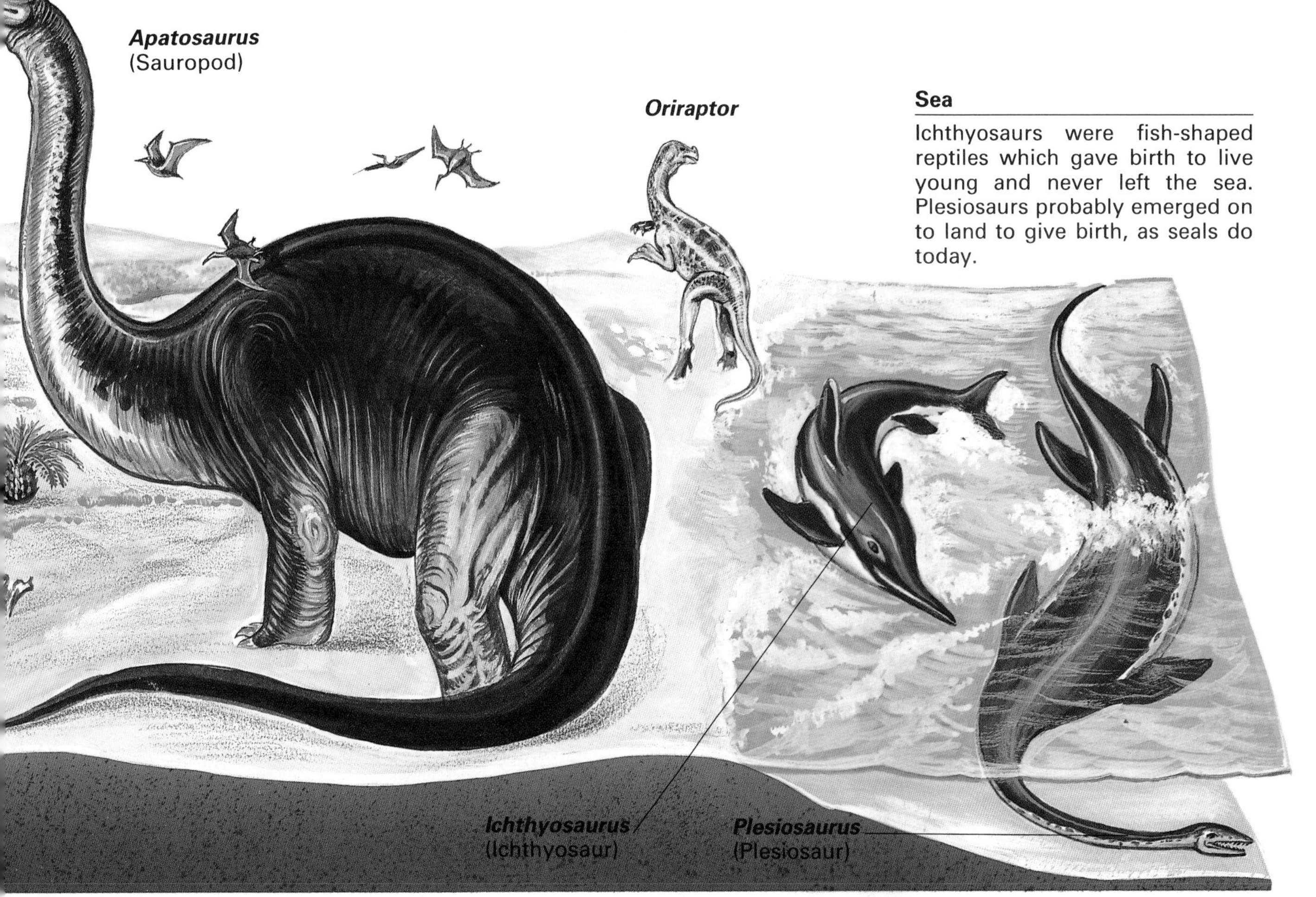

Sea

Ichthyosaurs were fish-shaped reptiles which gave birth to live young and never left the sea. Plesiosaurs probably emerged on to land to give birth, as seals do today.

CONQUERING THE AIR

Pterosaurs began to evolve about 220 million years ago, during a period known as the **Triassic**. They continued to flourish during the **Jurassic** and **Cretaceous**, and became extinct about 65 million years ago. Birds evolved about 150 million years ago, during the **Jurassic** period.

Flight has been "invented" four separate times: by insects, pterosaurs – extinct flying reptiles, birds and bats. No other animals can truly fly, although some, such as today's flying squirrels, can glide. Gliding animals float downward from one tree to another using some sort of "parachute" – usually a membrane between the front and hind limbs – to slow the fall and control where they land. The true fliers must have begun as gliders, but they went on to develop flapping wings from their original "parachutes." When membranes were replaced by separate feathered wings, the first bird had arrived.

Flying reptiles

At one time it was thought that the pterosaurs could only glide, but it is now clear that they were true fliers. One "finger" on each forelimb was enormously elongated to support the wing along its front edge. The later pterosaurs show many interesting adaptations for flight. Like modern birds, some had no teeth, a saving in weight that is useful to flying animals because teeth are very heavy. Pterosaur bones were hollow to make them extra light, and the lungs had extensions that went into the bones. Flying birds have similar sacs, which help to provide plenty of oxygen for the strenuous work of flying.

Some pterosaurs lived on fish, and one unusual species probably used its sieve-like beak to take plankton (a mass of tiny floating creatures) from the surface waters of the ocean. The smallest pterosaurs, which were no bigger than sparrows, ate insects. The largest species, *Quetzalcoatlus*, had a wingspan of 15m (49 ft) and probably fed on dead animals.

Early birds

A fossil known as *Archaeopteryx* shows how birds may have evolved from dinosaurs. The skeleton of *Archaeopteryx* is very like those of some small dinosaurs that walked on their hind legs, and there is little doubt that it was closely related to them. But *Archaeopteryx* had feathers – certain fossils show them very clearly. It was on the way to becoming a bird but it still had many reptilian features, including teeth and a long bony tail. Birds' tails are just long feathers growing from a short bony stump – all that remains of the true tail. *Archaeopteryx* was a glider rather than a flier, and its wings were not as fully developed as those of modern birds.

Loss of flight

Although they began as flying animals, some birds evolved into large flightless forms. Some of these are still alive today, such as the ostrich. Others are now extinct, including the moas, which were plant-eaters more than 2.5 m (8 ft) tall, and *Diatryma*, a ferocious predator that evolved after the dinosaurs died out. Because there were few large predators around, there were plenty of opportunities for an animal such as *Diatryma* to prey on small mammals.

Although an ostrich cannot fly, it can run fast.

THE FIRST MAMMALS

The first mammal-like reptiles appeared in the **Carboniferous** period, about 300 million years ago. They were the dominant land animals during the **Permian** and **Triassic**, between about 300 and 200 million years ago. The first true mammals appeared about 200 million years ago.

The mammals evolved from a group of reptiles that were very successful long before the dinosaurs even appeared. These "mammal-like reptiles" flourished for a hundred million years, but then most became extinct. Of those that were left, some gradually evolved into the mammals.

Although the first true mammals appeared at about the same time as the first dinosaurs, for 150 million years they remained small, nocturnal animals. Only when the dinosaurs died out did they get a chance to develop into new and larger forms.

Three types of mammals

There are three types of mammals, each with a different way of producing young. They all feed their young on milk. The placental mammals form the largest group – and the one to which humans belong. Their young develop inside the mother's body, nourished through a special organ, the placenta. A second group, the marsupials, also keep their young inside their bodies, but for a much shorter time. They give birth to tiny, grub-like young that complete their development in a pouch in the mother's body, where they feed on milk. Most marsupials are found in Australia. The third group of mammals, the monotremes, are the most primitive. They lay leathery-shelled eggs just as their reptile ancestors did. The small, grub-like young feed on milk after they hatch. The earliest mammals probably reproduced in the same way, but more advanced mammals keep their eggs inside the body.

Lions are placental mammals.

The gray kangaroo is a marsupial.

The platypus is a monotreme.

Teeth and jaws

The fossils of mammal-like reptiles provide very good evidence for the theory of evolution because they show a gradual transition from one type to another over a period of 100 million years. One feature that changed was the jawbone. The most primitive mammal-like reptiles had five separate bones in each half of the lower jaw (1). In later fossils, four of these bones gradually became smaller (2), until in modern mammals there is only one main bone (3).

1.

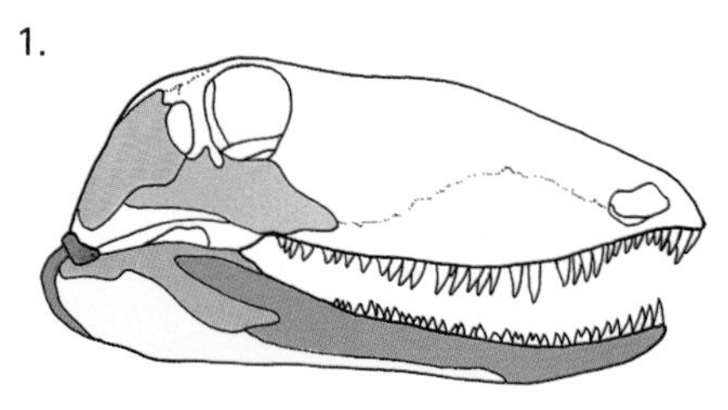

2.

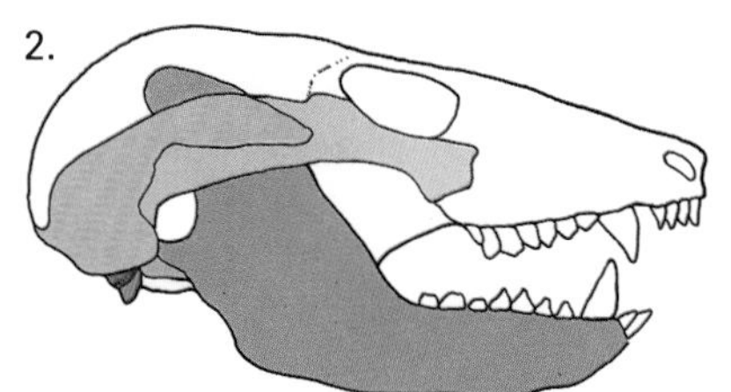

3.

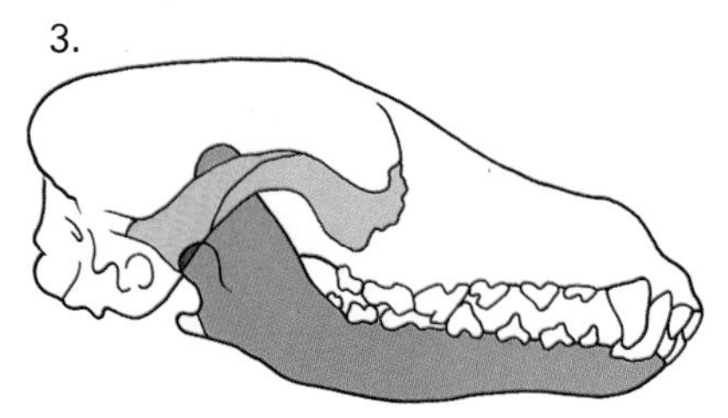

Out of the darkness

For 150 million years reptiles large and small dominated the land during the daytime, and mammals were probably restricted to a nocturnal existence. Most were shrew-like, the largest being no bigger than a cat. When the dinosaurs became extinct many opportunities opened up for the mammals. Their numbers multiplied enormously, and many new forms appeared. Some remained nocturnal, but many began to be active in the daytime. Over the next 50 million years there was also a steady increase in size in many mammal lines. One such group are the ancestors of the horse, which evolved from the dog-sized *Hyracotherium* to the large animal of today.

Modern horses are larger than their ancestors.

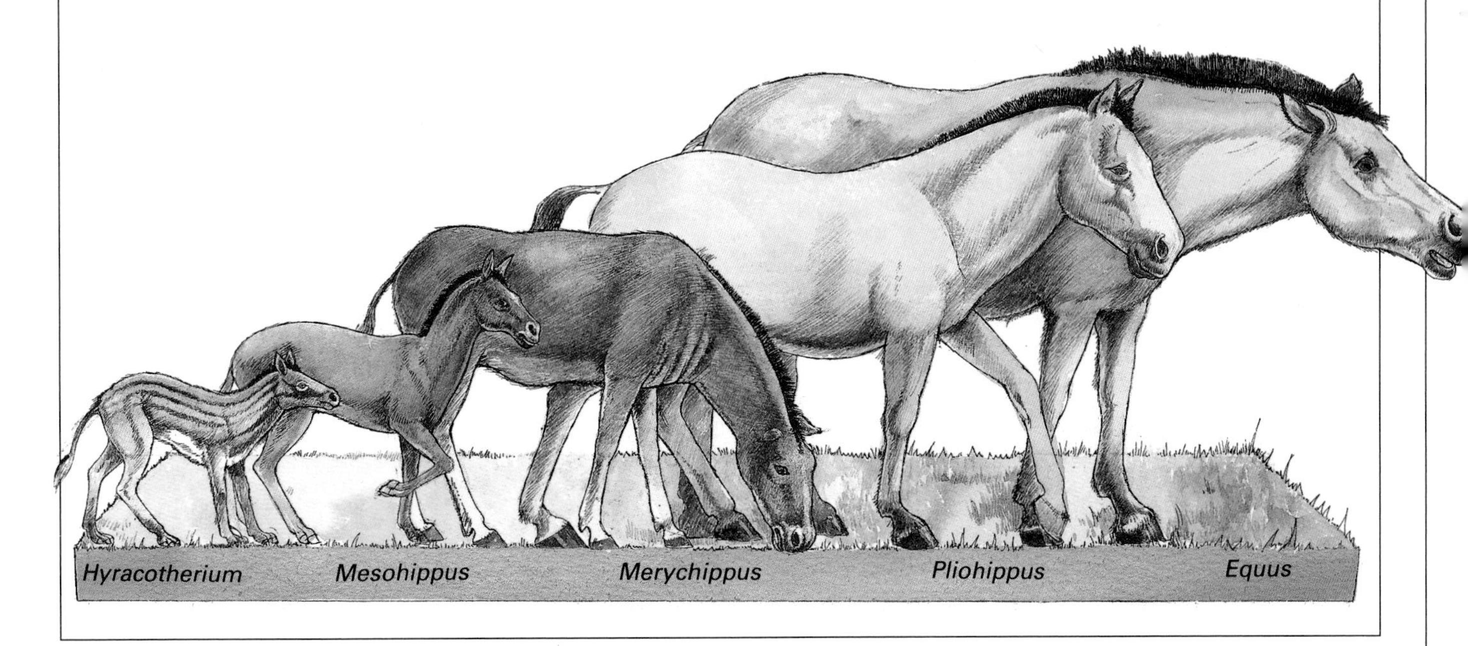

MAMMAL DIVERSITY

The Age of Mammals began 65 million years ago at the end of the **Cretaceous** period, when the dinosaurs died out. The largest living mammal at that time weighed about 2 kg (4.4lb). Today, the largest mammal is the blue whale, which can weigh 200 tons – 50,000 times as much.

The development of mammals has been greatly influenced by the changing position of the continents, which very slowly move about the Earth's surface. This is known as continental drift. Two hundred million years ago the continents had all collided to form a supercontinent called Pangaea. Over the past two hundred million years the different parts of Pangaea have been drifting steadily apart. At the same time the sea level has risen and fallen many times, so that separate land masses were sometimes linked by land and sometimes not.

Continental drift

200 million years ago

180 million years ago

65 million years ago

South American mammals

When South America became an island, mammal development was at an early stage. Some primitive placental mammals had already reached the continent, as had some marsupial carnivorous mammals, the *Borhyaenas*. During 10 million years of total isolation, many strange mammals evolved. These included placental mammals such as the plant-eater *Macrauchenia* and the giant ground sloth *Megatherium*. Many species were wiped out when the Americas reconnected.

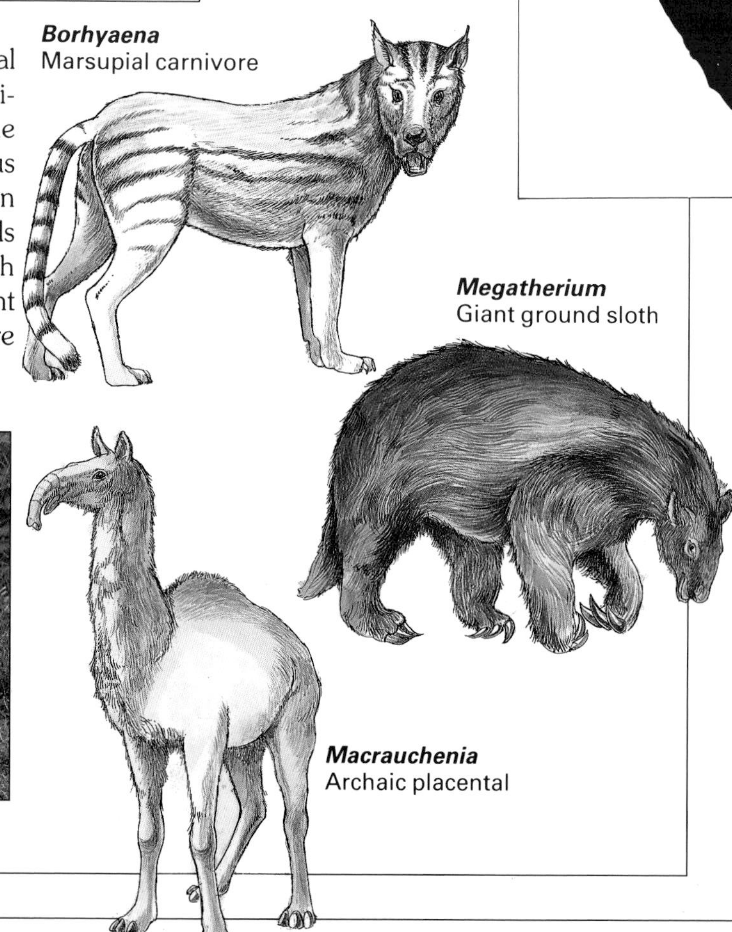

Borhyaena
Marsupial carnivore

Megatherium
Giant ground sloth

Macrauchenia
Archaic placental

A South American giant anteater and baby.

African mammals

Africa was virtually isolated from the rest of the world for long periods of time mainly because of higher sea levels. During these periods, animals such as *Arsinoitherium* arose, as did the elephants, which still survive today. Primates which had died out in most parts of the world survived in Africa and South America, and developed many new forms. In Africa they ultimately produced the Old World monkeys and apes. Humans also evolved in Africa.

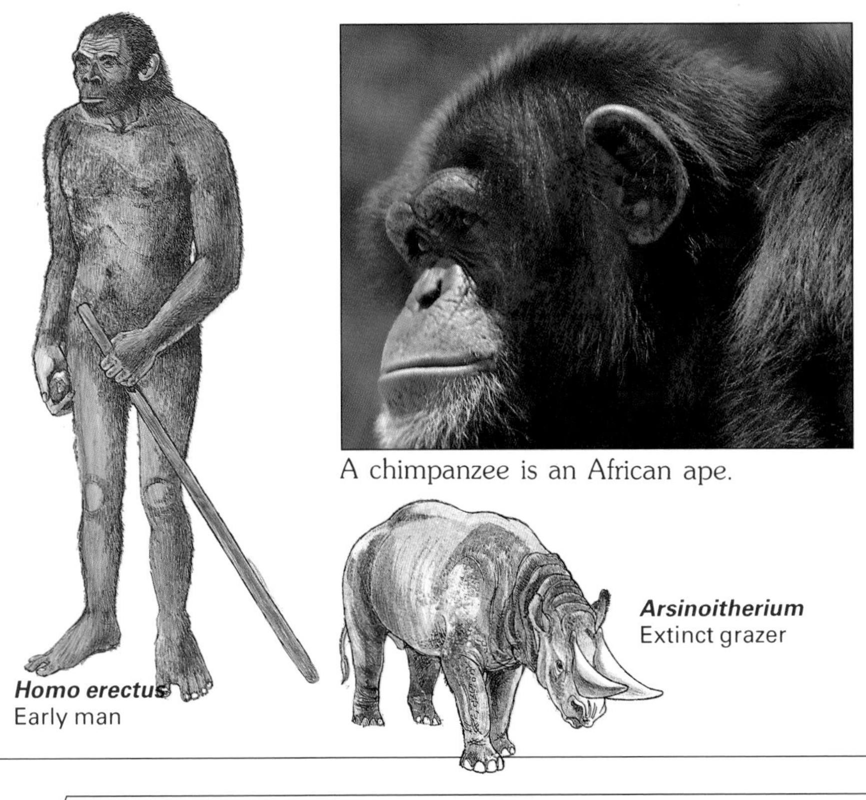

A chimpanzee is an African ape.

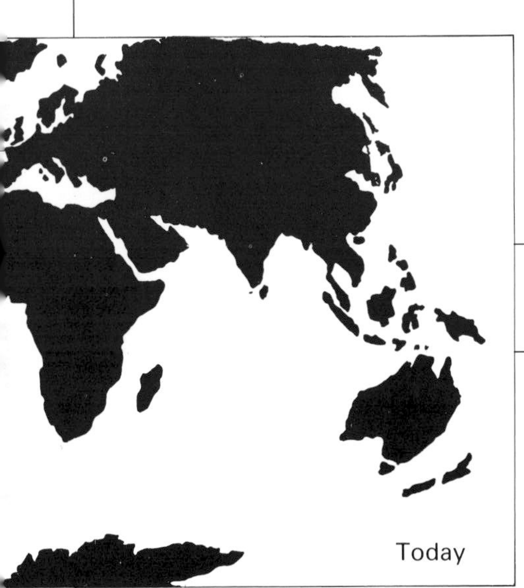

Australian mammals

Australia became cut off from the other continents during the early stages of mammal development. The marsupials had already arrived there, but not the placentals. So the marsupials survived and flourished in Australia, whereas in the rest of the world they lost out to the more advanced placentals. The only placental mammals to reach Australia were bats, which flew south from southeastern Asia much later in time. The egg-laying mammals, or monotremes, also survived in Australia, perhaps because there was less competition.

The koala is an Australian marsupial.

Flying mammals

Bats evolved more than 50 million years ago from small shrew-like ancestors. Like other flying animals, they probably began as gliders. As they developed, four of the "fingers" on each forelimb grew very long to support the membranous wing. The finger bones of a bat splay out like the spokes of an umbrella, giving its wing more support than that of a pterosaur, which had just one elongated finger bone along the leading edge of the wing.

By the time bats arrived, birds were well established. There were no new opportunities for daytime fliers, so the bats claimed the night.

Bats are mammals that have conquered the air.

Conquering the sea

Three separate groups of mammals have returned to the sea. The first, whales and dolphins, have completely transformed into aquatic animals. Instead of fur, they have a thick layer of fat under the skin, making them sleek and streamlined. But they must come to the surface to breathe air and can drown if trapped below water. They also die if stranded on a beach – crushed by their own weight when not in water. Dugongs and manatees, the second group, are also completely aquatic, but live in the quieter waters of estuaries and lagoons. The third group, seals and sealions, are far less specialized. They have fur and they return to the land to breed.

A manatee, or sea cow, "grazes" under water.

A dolphin is a mammal at home in the sea.

Extinctions

During the ice ages many very large animals roamed the Earth, including the imperial mammoth and the woolly rhinoceros. As the last ice age drew to a close, 10,000-15,000 years ago, most of these large animals became extinct. Overhunting by early man may have been the cause of these extinctions – they happened as human populations were expanding and developing new hunting skills. The saber-tooth cats died out about 11,000 years ago, perhaps because their prey animals (such as mammoths) were becoming scarce.

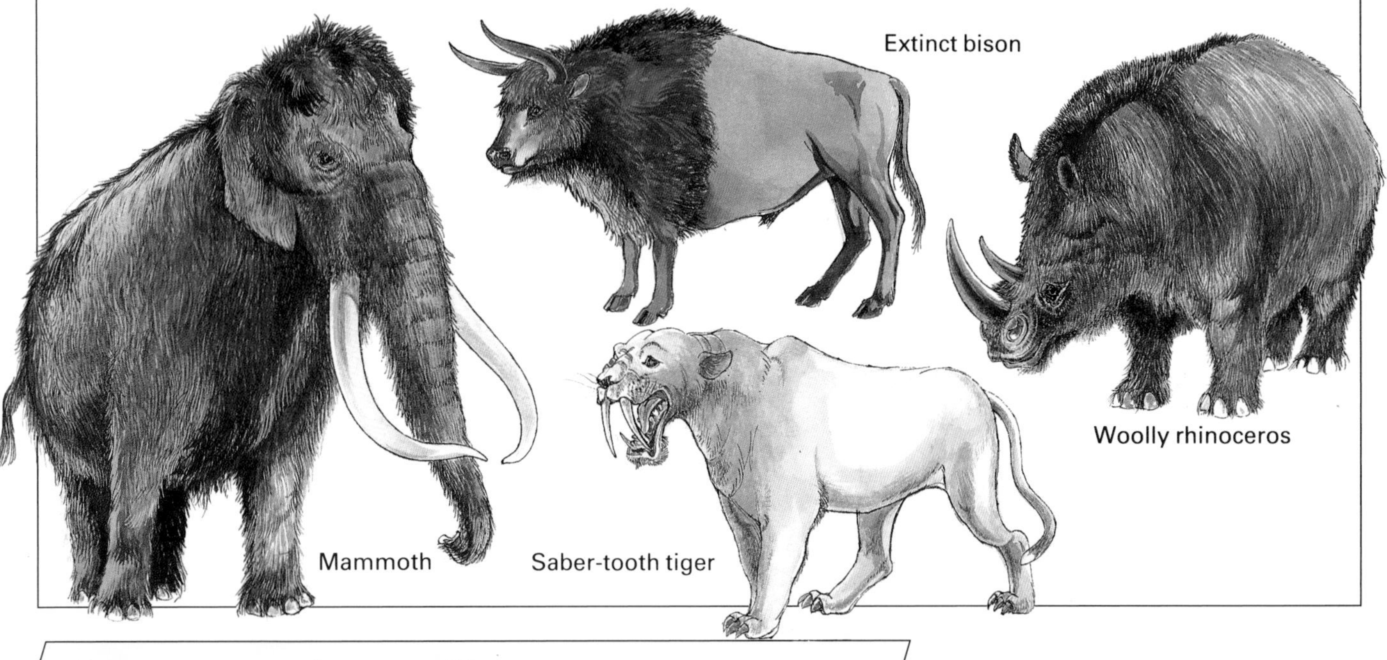

Large and small

Mammals are warm-blooded animals – they keep their body temperature at a constant level. To do this they have to eat extra food and turn some of the food energy into heat. Small animals lose a great deal of heat from their bodies to the air because the body is not very large in relation to its surface area. Tiny animals like shrews face the largest problem, and to stay alive they have to eat almost constantly. Very large animals such as elephants have the opposite problem – their heat-retaining body is huge. Having no fur helps them to keep cool.

The pygmy shrew is the smallest mammal.

The elephant is the largest land mammal.

HOW EVOLUTION WORKS

There are at least 4 million known species of living organisms alive on Earth today. They all use the same genetic code and their basic chemistry is the same. This suggests that they are all descended from a single ancestral species of bacterium that lived more than 3 billion years ago.

Evolution is a process of very gradual change – it is still happening today, but we do not notice it because the changes are far too slow. Sometimes, however, there is a sudden alteration in the environment that produces more rapid changes in animals or plants. During the last century, soot from factory chimneys blackened tree trunks in industrial areas. Moths that rested on the trees were originally pale and mottled to give camouflage against the lichens on the tree trunks – this protected them from insect-eating birds. But as the tree trunks turned black with soot, new black forms of the moth appeared. Today both the black (or melanic) moths and the original pale forms are found, but the black ones are becoming less common as air pollution is controlled.

Geographical barriers are important in evolution because they can help to form new species. If a small group of animals, or a few plant seeds, somehow cross the barrier, they become isolated from the parent species and may give rise to new forms.

Deserts

Over thousands of years, an area that was once green can turn to desert. A species that cannot cross arid lands may then be split into two groups, separated by desert. In time they may develop into separate species.

Rivers and mountains

Rivers form a barrier to some small animals. High mountains are a barrier to many animals and plants. In exceptional circumstances, some may cross these barriers and set up breeding populations, which can eventually become new species.

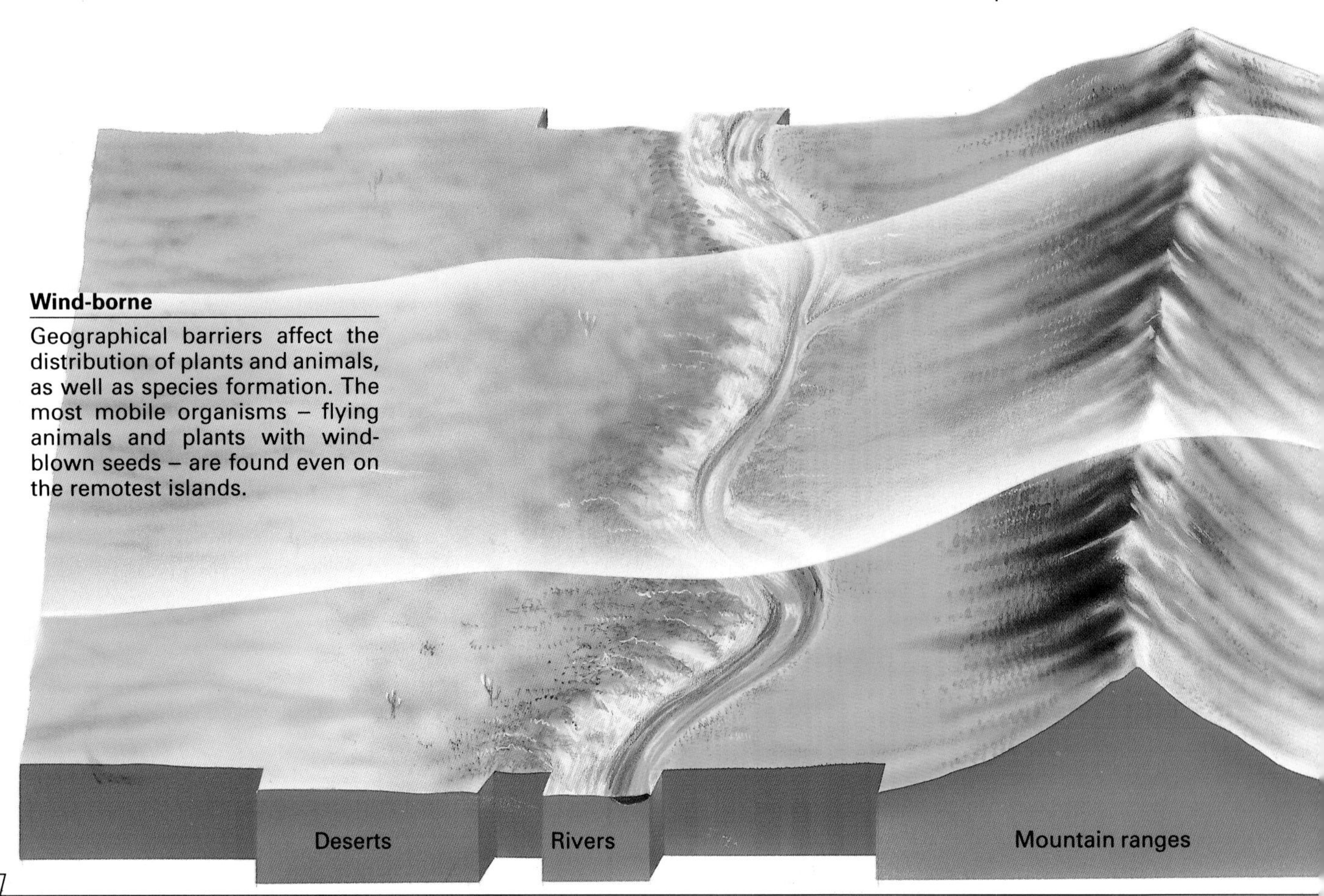

Wind-borne

Geographical barriers affect the distribution of plants and animals, as well as species formation. The most mobile organisms – flying animals and plants with wind-blown seeds – are found even on the remotest islands.

Forming new species

For evolution to happen a new species must be formed from an old one – the parent species. The formation of a new species can happen in several ways. Animals may fail to breed with each other simply because they no longer look the same, or behave the same way during courtship. Or they might fail to breed because they have become cut off from the parent species by a geographical barrier. Once this has happened and the two groups stop interbreeding, they evolve in different ways.

There are several species of giant tortoise found in the Galapagos.

Oceans

Vast expanses of sea are the greatest barrier of all. It is usually only flying animals that can cross them, although small mammals sometimes float on pieces of driftwood, reaching islands near the mainland.

Evolution on islands

Many islands, such as the Galapagos, have been formed by undersea volcanoes. These islands start out completely barren, with no life forms at all. The few species to arrive have the island to themselves and often evolve in unusual ways as a result. Other types of island, such as Madagascar, form by breaking away from a mainland. Because they are isolated, they too may develop unusual life forms. Ancient species, such as lemurs, can survive on these island refuges because they do not have to compete with more advanced animals.

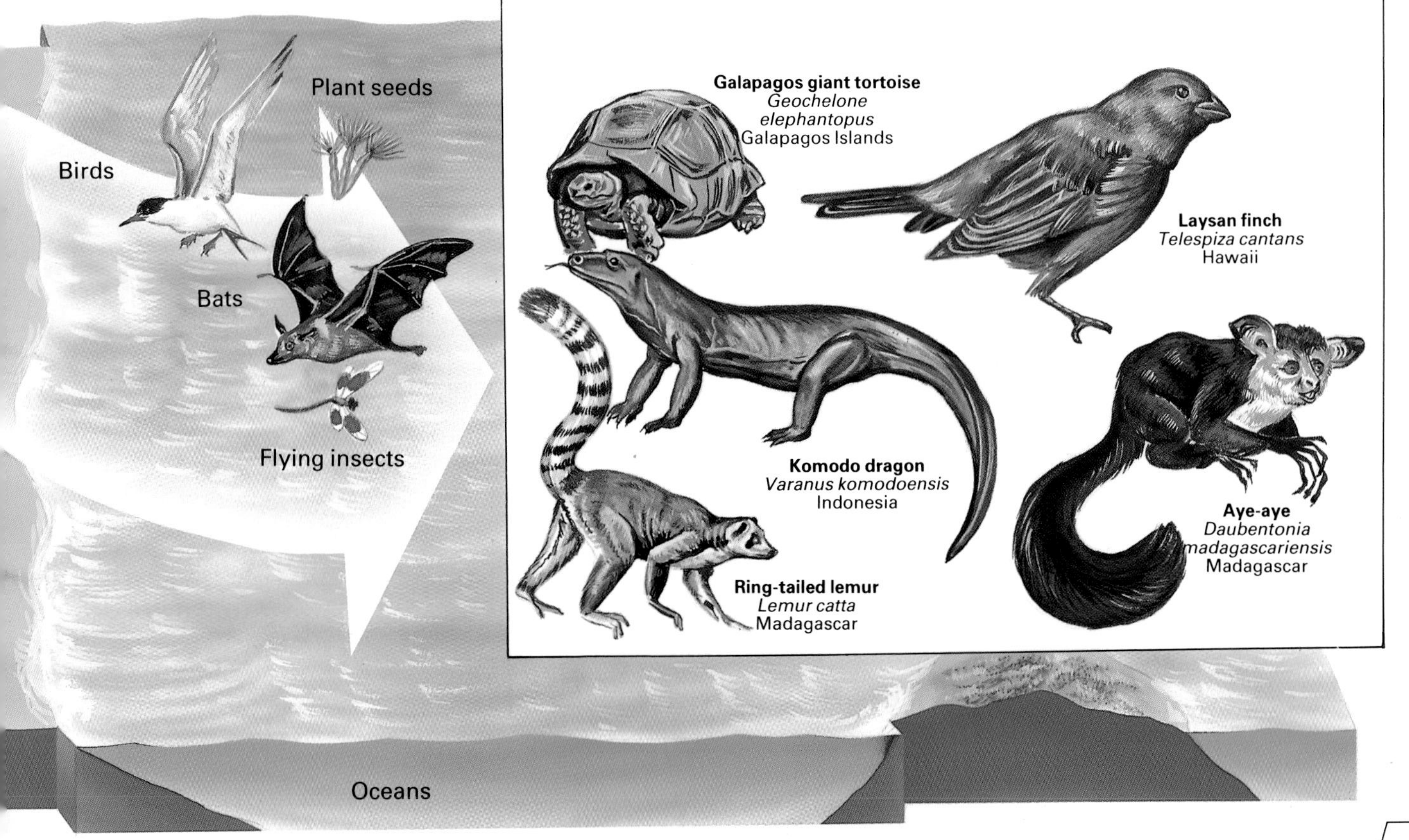

Natural selection

The main driving force behind evolution is natural selection, or "survival of the fittest." Several factors affect natural selection. For instance, no two members of a species are ever exactly the same. Those that survive may do so because they are healthier or stronger than others of their species. Or perhaps they have better camouflage – a disguise of color or patterns. All of these characteristics will be passed on to their young. This process is called inheritance (see page 28).

The result of natural selection is the development of individuals (and ultimately species) that are more resistant to disease, that are stronger and that are better adapted to their particular environments. With time, the characteristics of the individuals that do not survive may die out.

Wild boar piglets have protective coloration.

A giraffe has a tough mouth to eat thorny acacia.

Mutations

From time to time, random changes called mutations take place in the genes of plants and animals giving rise to modified species. They have unpredictable effects – they can produce beneficial changes, harmful changes or no visible change at all. Most harmful mutations result in the death of the offspring before it is born.

A common less harmful mutation is albinism, in which all dark pigment (the coloring in the cells of plants or animals) is lost. The opposite mutation produces melanism (an unusual amount of dark pigmentation), as seen in dark melanic moths (see page 24).

An albino blackbuck is a rare mutation.

Adaptations

One effect of natural selection is to favor adaptation: the fitting of an animal or plant to its way of life. Animals that live in deserts, for example, have undergone selection for features that save water. Camels have humps that can yield moisture. Many desert plants have a waxy coating on their leaves to reduce water loss.

As well as having adaptations to the physical environment, organisms must adapt to each other. This may mean developing long legs to escape from predators, or a poisonous sting to paralyze prey. In the case of flowering plants, their brightly colored flowers, well-supplied with nectar, are an adaptation to tempt flying insects such as bees to visit them and thus transfer pollen to another flower.

A camel's humps store fat.

Nectar attracts pollinating bees to flowers

Isolation

Once a new species has formed, it does not go back to breeding with the parent species, or with any other species. It now has its own adaptations which suit its particular way of life – individuals looking for a mate will not benefit by diluting their genes with those of another species. (The only important exception to this is in plants, where hybrids between species sometimes occur). To avoid interbreeding with other species, "isolating mechanisms" have evolved. For many animals, these isolating mechanisms involve courtship rituals: songs, calls and various displays that are unique to their species.

A female blackcap calling from its perch.

GENETICS AND HEREDITY

Inheritance – a major factor in natural selection and therefore in evolution – takes place by means of genes. Any offspring has a combination of characteristics inherited through a mixture of genes from its parents. The basic genetic material is DNA, which is found coiled up very tightly inside structures called chromosomes in the nucleus of the cell.

We all know that certain characteristics "run in families" – children tend to look like their parents. What we are seeing is the work of the genes – packages of inherited information that play such an important part in natural selection. Genes tell us how to develop, keep all our bodily functions running smoothly, and determine family characteristics such as hair color and eye color. All animal and plant genes are made up of DNA. They are strung along in chromosomes, strands of material located in the nucleus of every cell. When cells divide to form new cells, the chromosomes, genes and DNA also have to be reproduced. It is this ability to copy itself that makes DNA the key substance of heredity.

The diagram explains how DNA controls the assembly of chains of amino acids to form proteins. Only a very small part of a DNA molecule is shown.

DNA

DNA consists of two long strands twined together like a twisted rope ladder. The information in the DNA is used to make other long-chain molecules, called polypeptides. These are then assembled into proteins. The long strand of DNA in each chromosome represents thousands of genes. A gene is simply a length of the DNA molecule that controls the formation of a particular polypeptide chain.

Messenger RNA

DNA (1) sends its instructions out of the nucleus in a molecule called messenger RNA (2). This is a chain of nucleotides which are a copy of a length of the DNA that corresponds to a single gene.

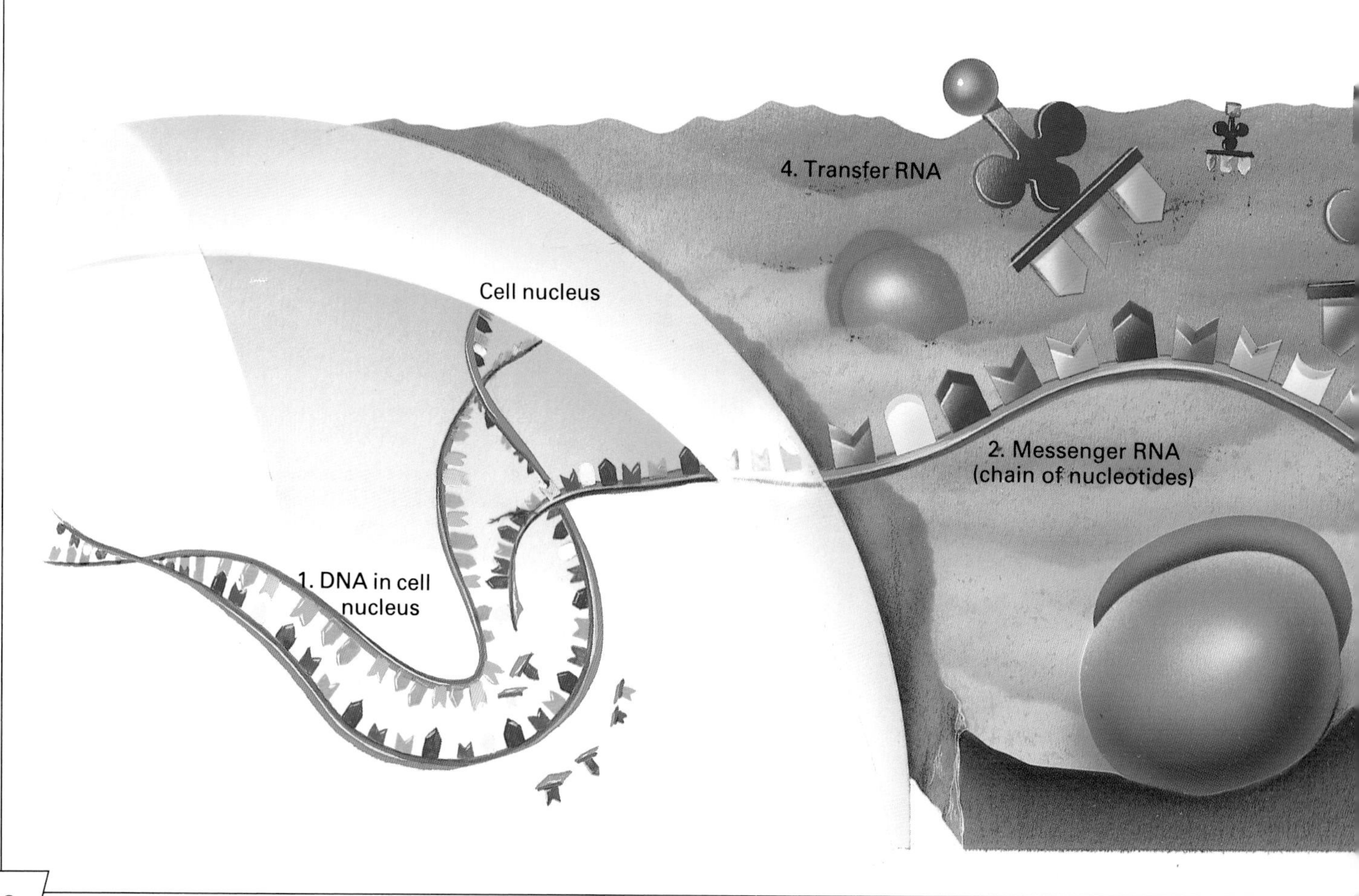

Enzymes in action

Some of the proteins produced under the control of DNA make up bone, skin, muscle or other structural components. Many others act as enzymes – catalysts that control the living processes of the body. Enzymes are responsible for making all the other constituents of the body, such as fats, sugars and pigments. So by controlling the enzymes, DNA controls the whole body.

Mutations – changes in the DNA – can change the structure of enzymes so that they no longer work. In albinos one of the enzymes that makes the dark pigment melanin is affected. A slightly different mutation produces a Siamese cat's coloring. A melanin-making enzyme has become sensitive to heat – it can make the dark pigment only in the colder areas of the body.

All Siamese cats have blue eyes.

Transfer RNA

Outside the nucleus, but still inside the cell, ribosomes (3) help to translate the instructions in the messenger RNA. Starting at one end of the RNA chain they "read" the nucleotides in groups of three. Each group is called a codon, and every codon corresponds to a particular amino acid, which is carried into place by transfer RNA (4). This happens because each transfer RNA also has a set of three nucleotides called an anticodon which pairs with a particular codon. The other end of the transfer RNA molecule combines with a particular amino acid (5), so the codon is always translated into the correct amino acid. When the amino acids are put together in the order shown by the messenger RNA's codons, they produce the correct polypeptide chain (6), which in turn builds up into a protein. Once the transfer RNA has delivered its amino acid, it moves away (7).

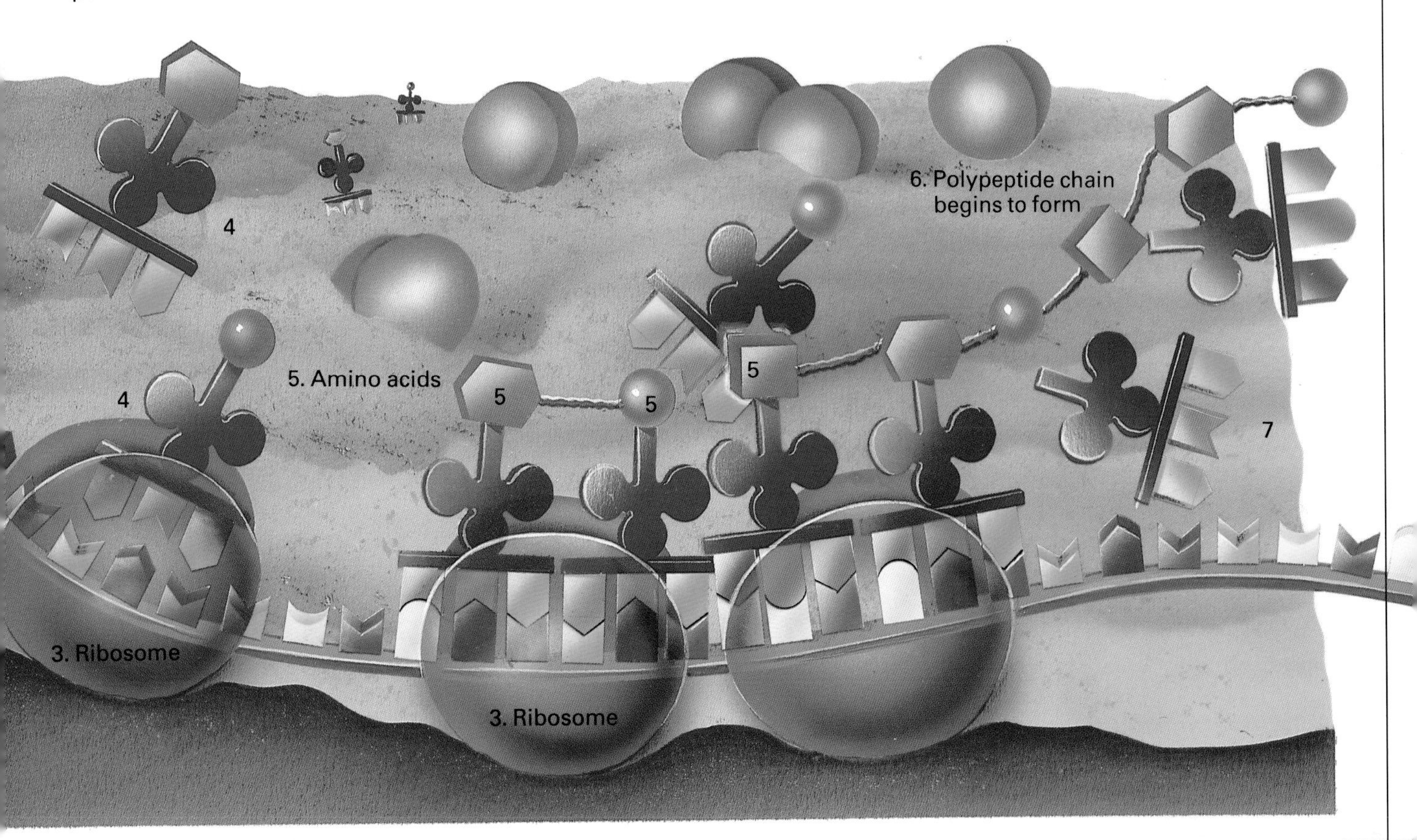

Parents and offspring

In most plants and animals, the cell's nucleus has two versions of each chromosome – one inherited from the mother and one from the father. The animal or plant therefore has two versions of each gene. Sometimes both of these genes are active, but often it is only one – called the dominant gene – that produces noticeable effects. The gene whose effect is masked is called recessive. This is why characteristics sometimes skip generations – they may be shared by children and grandparents but absent from the parents. Such characteristics are coded for by recessive genes. One parent must have carried the recessive gene, but because he or she also carried the dominant gene its effects were not noticeable. Only when an individual has two copies of the recessive gene are its effects seen. In the mouse, for example, black coloration is dominant and brown is recessive. The diagram shows the result of crossing a black and brown mouse, and then crossing one of the daughters with one of the sons. All mice in the first generation are black because the gene for black is dominant. But in the second generation, two recessive (brown) genes come together and produce one brown mouse among four offspring.

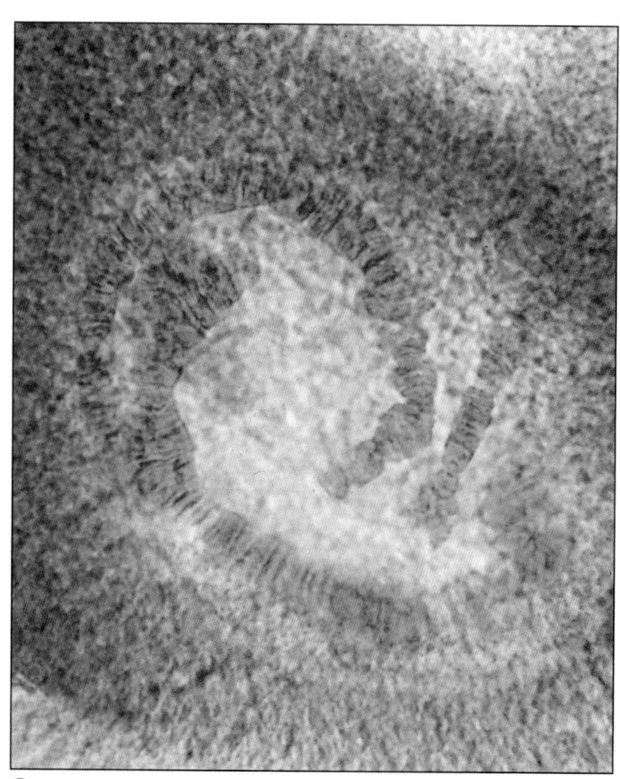

Gene-carrying chromosome (blue) from a fly

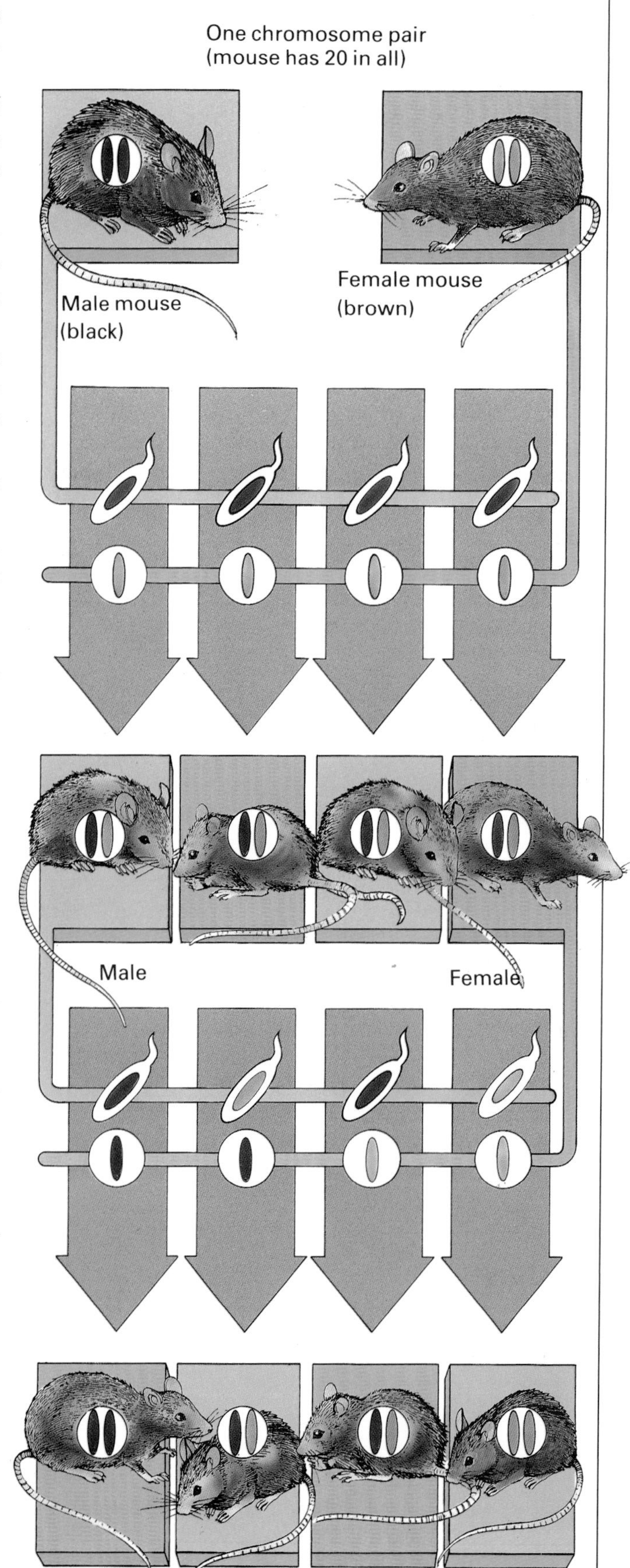

Not all in the genes

All living things are affected by their environment, which combines with the total action of their genes to produce their size and appearance. Animals of one species tend to look alike because animals need to move around and this places constraints on their size and shape. Species of plants are far more affected by their environment because they remain in one place. Plants may grow to extraordinary heights in search of sunlight, or may become misshapen by the wind – characteristics that will not be passed on in the genes.

The shape of trees can be formed by the wind.

Artificial selection

When farming began, about 10,000 years ago, humans took certain plants and animals from the wild and started using them for food. These early farmers always chose the best animals to keep in their flocks, and took seed from the best plants to sow the following year. The animals and plants gradually changed as a result of this process, known as artificial selection. Today, plant and animal breeders still practice artificial selection, but they now base their breeding on an understanding of genetics. Artificial selection works in a similar way to natural selection, but it is faster and less haphazard because breeders know exactly what they want and discard all but the best. Dogs are a good example of what artificial selection can achieve. Starting with one wild species – the wolf – it has produced dogs in a huge range of shapes and sizes.

All these vegetables are types of cabbage.

EVIDENCE FOR EVOLUTION

It takes 15,000 years to deposit a meter (3 ft) of sediment in an average river or sea. Scientists estimate that it may take at least 300,000 generations for a new species to evolve. In the case of the Hawaiian finches, it took less than 5 million years to produce 28 different species from a single ancestor.

No one can prove evolution because it takes place so slowly, and many of the events that interest us – such as the evolution of birds – happened so long ago. In the same way, we cannot *prove* scientific theories about how the world began or how an atom is made up – but we accept, from the great weight of evidence, that the theories are correct. In the case of evolution, there is just as much good evidence, and it comes from a variety of sources. The accumulated evidence is so strong that biologists accept evolution as a fact and see it as a continuing process.

The fossil record

The fossil record stretches back for millions of years, but the oldest rocks contain only very simple creatures such as bacteria, worms and jellyfish. More complex organisms, like birds, mammals and flowering plants, first appear in much later rocks. This is regarded as evidence that life began with simple forms and gradually developed into more complex ones. In some cases there are clear sequences of fossils that show how evolution occurred. A good example is the transition from reptiles to mammals, for which there is a known sequence of intermediate fossils. Such sequences are unusual, however, and the chance of an animal or plant dying in the right situation to be preserved is extremely small.

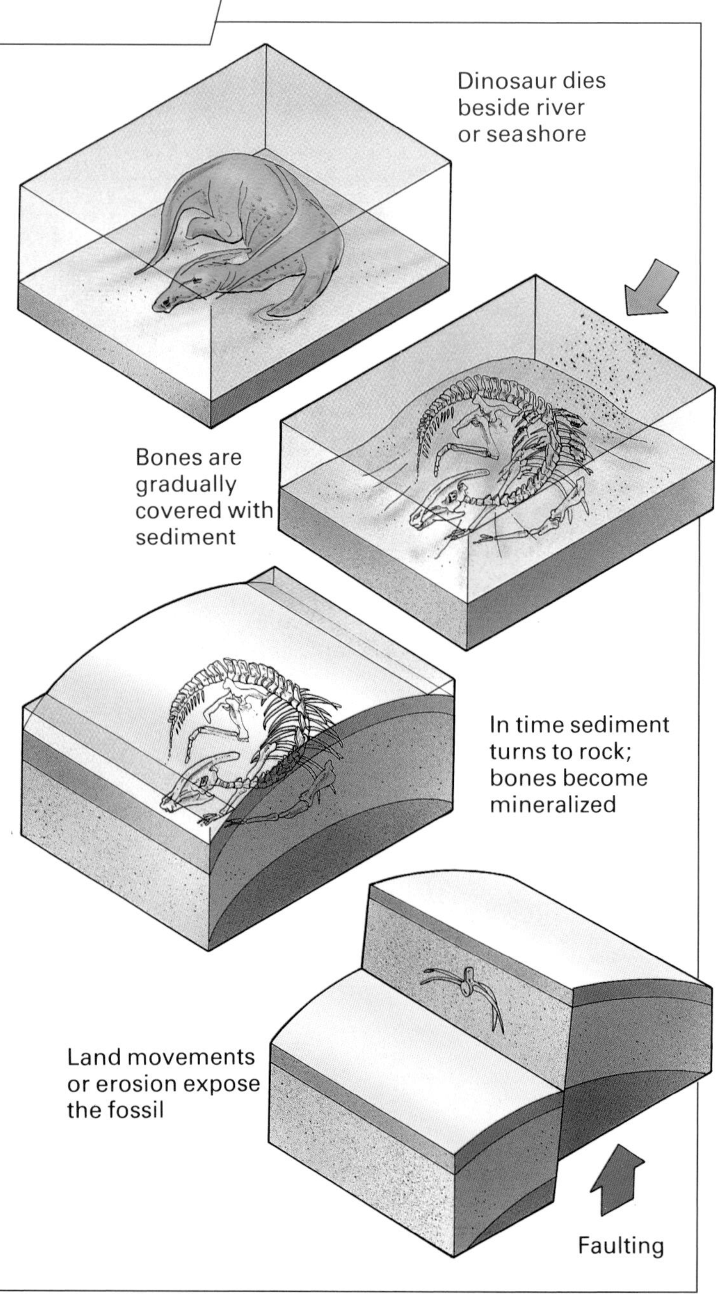

Fossil fern leaves in limestone rock

Shared ancestors

Natural selection works by individual changes on an existing body plan – it cannot redesign the basic structure. So related animals or plants should share the same basic structure. The fossil evidence shows that mammals, birds, reptiles and amphibians are all descended from the same ancestors.

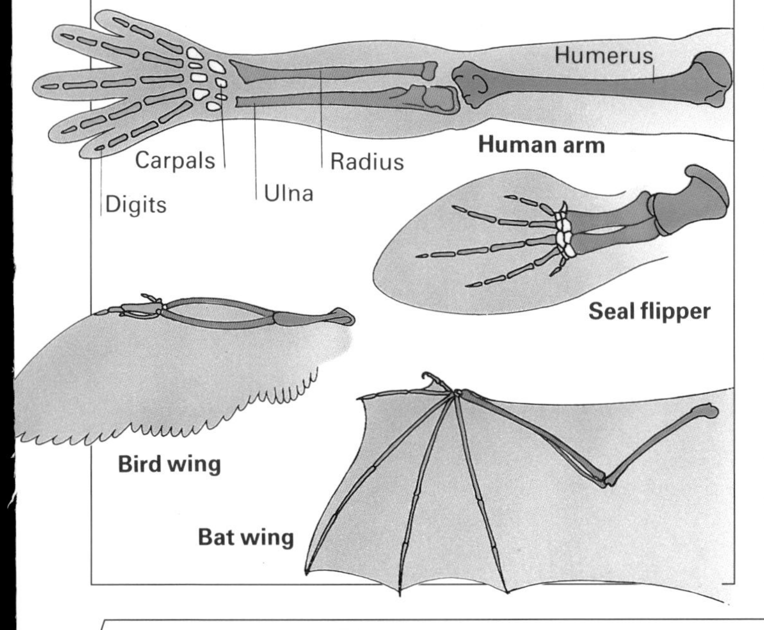

Embryos

A look at the embryonic and larval stages of some species can provide evidence for evolution. The larvae of barnacles look like water-fleas, while the adult barnacle develops into a fully shelled crustacean that fixes itself to rocks and ships' hulls. This suggests that barnacles originally evolved from creatures like water-fleas that swam around freely.

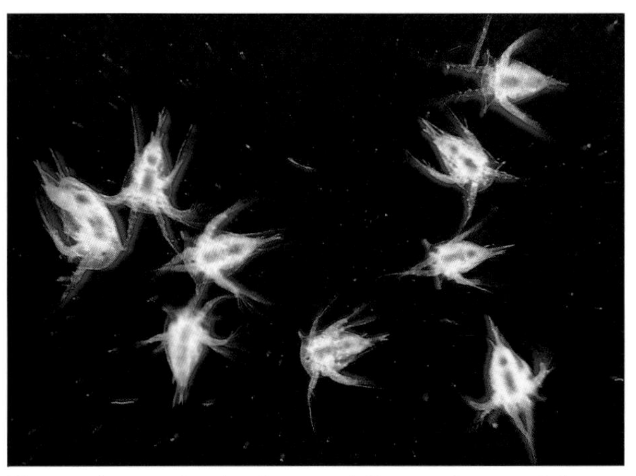

Young barnacles look very different from adults.

Living evidence

Important evidence for evolution comes from oceanic islands. The few species that reached these islands had a perfect opportunity to diversify. In the Galapagos, for example, one species of bunting evolved into 13 different species, the Galapagos finches. Each exploits a different type of food, some taking seeds, others insects. Every species has a different type of beak, suited to its diet, and the birds use beak size and shape to recognize their own species during courtship. In a similar process in Hawaii, a single species of finch or honeycreeper gave rise to 28 different species, with a wide range of feeding habits and beaks.

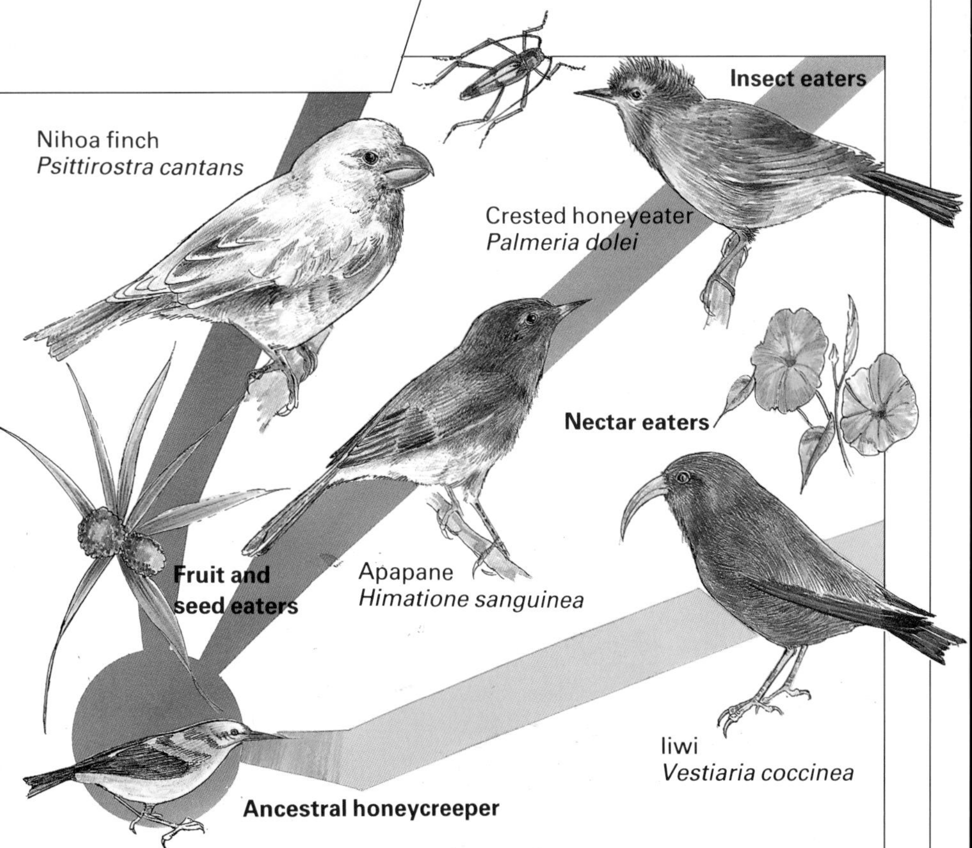

EVOLUTIONARY TIME CHART

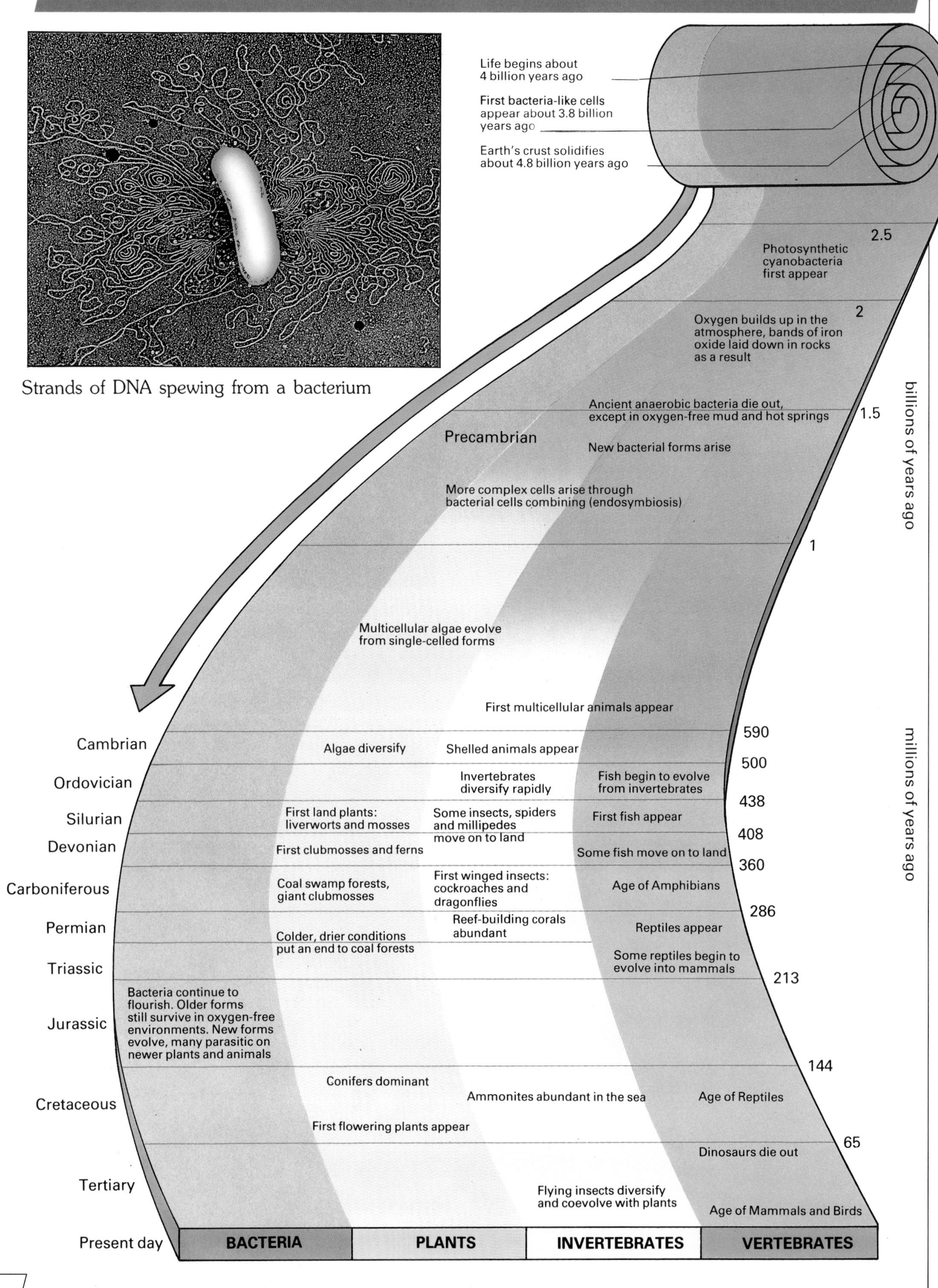

Strands of DNA spewing from a bacterium

GLOSSARY

algae group of plants that includes seaweeds and many pondweeds, as well as many single-celled plants. Some live as a powdery layer on tree trunks and a few live inside other organisms, but most algae are confined to water.

amino acids chief components of proteins, *see* polypeptide.

cell basic building block of most living things (except viruses and many fungi). Cells are distinct microscopic units of living matter, surrounded by a membrane and containing jelly-like protoplasm.

chloroplast egg-shaped structure inside the cells of plants, which carries out photosynthesis.

chromosome thread-like or sausage-shaped structure, in the nucleus of a cell, which carries genes.

cyanobacteria relatives of bacteria that photosynthesize and produce oxygen in the process. They have more complex cells than bacteria, with many infolded membranes that carry out the photosynthesis.

DNA deoxyribonucleic acid, the hereditary material that makes up genes.

dominant of two genes for the same characteristic (such as eye color), the dominant gene is the one that is expressed. The one that is not is called recessive.

enzyme specialized protein molecule that helps chemical reactions happen in living organisms. By controlling such reactions, enzymes control growth and development as well as the day-to-day running of the body, the breakdown of food for energy, and the repair of damaged tissues. Each reaction has its own particular enzyme.

eukaryote organism with one or more complex cells containing mitochondria and other membrane-bound structures. DNA is arranged in chromosomes in eukaryotes, and the chromosomes are inside the cell nucleus. All living organisms except bacteria and cyanobacteria are eukaryotes.

fossil remains of a living organism, preserved in rocks.

gene unit of hereditary information which passes characteristics from parents to offspring. It corresponds to a length of DNA that forms part of a chromosome.

hybrid offspring of a cross between two different species, or between two inbred lines of the same species.

invertebrate animal without a spine or backbone.

larva young of an animal such as an insect or amphibian, whose form is very different from that of the adult.

multicellular made up of many cells.

mutation change in the DNA making up a gene.

nucleotide basic structural unit of DNA and RNA.

nucleus central "core" of a complex cell (eukaryote). Enclosed by a double membrane, contains chromosomes.

parasite organism that lives in or on another one and depends on it for all its food.

photosynthesis process that uses light, carbon dioxide and water to make food. It occurs in the green parts of plants, in cyanobacteria and in some other bacteria.

pollen fine powder-like material produced by flowering plants. It contains the male gametes (sperm), and when transferred to another flower fertilizes the egg cells so that the plant can form seeds.

polypeptide long-chain molecule made up of strings of amino acids. Polypeptide chains make up proteins.

primitive animal that resembles its distant ancestors, not necessarily one that is simple in structure or undeveloped.

prokaryote simple cell with no nucleus or mitochondria. Bacteria and cyanobacteria are prokaryotes; more complex cells are eukaryotes.

protein molecule made up of one or more polypeptide chains. Proteins are important constituents of blood, bone, hair, muscle and skin. They also act as enzymes.

protozoa single-celled organisms that have no rigid cell wall but can (in most species) move about. They have complex (eukaryote) cells, and a few have chloroplasts.

recessive of two genes for the same characteristic, one that is not expressed unless paired with another recessive gene.

RNA ribonucleic acid, a molecule that is chemically similar to DNA but has different roles within a cell.

species group of living organisms that can interbreed, and which do not normally breed with organisms of another species. Members of a species usually look similar.

variety group of organisms within a species that share one or more distinctive characteristics, but are able to breed with others of the same species.

vertebrate animal with a backbone. Vertebrates include fish, amphibians, reptiles, birds and mammals.

unicellular consisting of just one cell.

INDEX

All entries in bold are found in the Glossary

Photographic Credits
Cover and page 6 (both): Science Photo Library; intro page: Bruce Coleman; pages 5, 7 (bottom), 8, 10, 11 (top left, bottom left and bottom right), 12, 13 (both), 17, 18 (top right), 21 (bottom), 22 (all), 26 (bottom), 27 (bottom) and 29: Planet Earth/Seaphot; pages 7 (top), 21 (top), 23 (right) and 27 (top left and top right): Robert Harding Library; pages 9 (both), 11 (top right), 18 (left), 23 (left), 31 and 32: Bruce Coleman; pages 18 (bottom right), 19, 20, 26 (top left) and 28: Survival; page 26 (top right): Hutchison Library; page 33 (top): L. Gamlin; back cover: Science Photo Library.